Women in Agriculture

WOMEN IN AGRICULTURE

Professionalizing Rural Life in North America and Europe, 1880–1965

LINDA M. AMBROSE AND JOAN M. JENSEN, EDS.

University of Iowa Press, Iowa City

University of Iowa Press, Iowa City 52242

www.uipress.uiowa.edu
Printed in the United States of America

Design by Kristina Kachele Design, llc

The University of Iowa Press is a member of Green Press Initiative and is committed to preserving natural resources.

Printed on acid-free paper

Library of Congress Cataloging-in-Publication Data
Names: Ambrose, Linda McGuire, 1960– editor. | Jensen, Joan M., editor.
Title: Women in agriculture : professionalizing rural life in North America and Europe, 1880–1965 / Linda M. Ambrose and Joan M. Jensen, eds.
Description: Iowa City : University of Iowa Press, [2017] | Includes bibliographical references and index.
Identifiers: LCCN 2016033496 | ISBN 978-1-60938-472-2 (pbk) | ISBN 978-1-60938-473-9 (ebk)
Subjects: LCSH: Women in agriculture—North America—History—19th century. | Women in agriculture—North America—History—20th century. | Women in agriculture—Europe—History—19th century. | Women in agriculture—Europe—History—20th century.
Classification: LCC HD6077 .W663 2017 | DDC 331.4/8309409041—dc23
LC record available at https://lccn.loc.gov/2016033496

The Rural Women Studies Association (RWSA) has provided a supportive culture for our research into rural women and food studies. This collection is only one sample of the breadth of that ongoing research. We dedicate this book to those who collectively have created that research and who sustain it through RWSA.

Contents

Preface

Over the years, each of the contributors to *Women in Agriculture* has written about the history of rural women. Those writings have included women's relationship to food, but food has never been our primary focus. Instead, food has been one of many threads woven into the fabric of the lives of the women we study. In the present collection, we bring those threads to the surface with stories about women whose professional lives were closely tied to food because of their work as educators, experts, and extension workers. The years from 1880 to 1965 were a time of great change in agriculture as a result of mechanization and improved transportation and communication, and all the women whose stories we tell shared a commitment to improving the lives of ordinary people.

The recipes for rural life that these professional women created form a part of the rapidly developing academic field of food studies. Like women's studies in the 1970s, food studies has entered universities in response to student demand, through an academic cohort steeped in popular cultural concerns surrounding food security, sovereignty, and biopiracy. In fact, these concerns have always been a part of the academic study of food, embedded in disciplines such as agricultural and rural history, agricultural economics, ethnobotany, and rural sociology. The development of food studies has encouraged us to look more carefully at this past research, to ask new questions about women's historic roles in the production and marketing of food, and to employ cross-disciplinary methods in finding answers to those questions.

Like others who have studied food, we are situated in specific food cultures that can be divided into those based on wheat, rice, or corn. Given those three options, we definitely come out of wheat cultures, in which bread was the basic staple. Because memories of food are a part of those cultures and a spur to studying that process, our memories of bread are part of the food culture we bring to this study. From Marcel Proust, who enshrined the French madeleine in *Remembrance of Things Past* to Diane Tye, who studied her mother's recipe cards in *Baking as Biography,* and Denise Chávez, who wrote a tribute to her parents in *A Taco Testimony: Meditations on Family, Food and Culture,* memoirs have enriched our experience with food. And so we begin with our own memories about bread.

One of my (Linda's) earliest food memories is walking up the lane to my rural home after school on "bread-baking day." Imagine my anticipation as the aroma of hot bread greeted me! I couldn't wait to enjoy a big, thick slice of it, then lick the melted butter from my fingers. In the late 1960s my mother baked bread once a week for our large family, producing an average of twenty loaves at a time. While I loved baking day, I remember feeling very self-conscious about taking clumsy homemade bread sandwiches to school. In fact, I felt jealous of my "town" friends and some farm friends too, who all seemed to be enjoying something I hardly ever tasted: white, "store-bought" bread. We called it "bakers' bread," and it was a very rare commodity in our home, reserved for special-occasion sandwiches for parties or funeral receptions. As a child, I always thought we ate homemade bread because my family couldn't afford to buy the commercially prepared kind. Many of my friends' mothers worked off the farm in jobs as nurses, clerical workers, or teachers while my mother "did not work," at least not for pay, and not off the farm. When I was young, I gave no thought to the superior taste or nutritional value of the different bread we ate compared to what my friends enjoyed. When I became a teenager, I enrolled in a 4-H club and learned how to make several different varieties of bread myself. Despite my efforts, none of my bread seemed to match the flavor or texture of my mother's. While my mom had no formal culinary training, when it came to home baking, in my opinion and that of others who tasted her bread, she was an expert.

My childhood memories of bread, I now realize, reflect a number of historical trends. First, there was the growing ubiquity of commercially produced foods and prepackaged cakes and pastries. As it became more difficult to live from farm incomes alone, more farm women took paid work off the farm and abandoned home baking for the efficiency and economy of so-called convenience foods. At the same time, professional home economists continued to demonstrate food preparation techniques in short courses given to community 4-H club leaders and to judge young rural girls like me on our mastery of baking and other homemaking techniques. Meanwhile, my "untrained" mother clearly excelled as a baker. Her expertise was grounded in experience, not formal qualifications.

I (Joan) have completely different memories of bread. My mother, who had grown up on a small Wisconsin farm and left for the city and eventually a middle-class life, hated that early life for its poverty and deprivation. While my grandmother had baked bread, my mother always gave us store-bought bread. Nor did she teach me to cook. She did, however, teach me to read. I still remember looking at the illustrations for the Grimm's fairy tales with which she taught me those reading skills and introduced me to the dogs with eyes as big as saucers. I must have been four when we sat together looking at and reading those books. After I left home at age twenty, I taught myself to cook from cookbooks.

My less affluent aunt, however, always baked bread, even though she worked away from home to supplement her husband's meager income. She was, in fact, a breadwinner as well as the bread maker. Years later, I reminded her daughter how I looked forward to her mother's homemade bread, how she served us thick slices of the bread slathered with jelly. My cousin replied, like my coeditor, that she had envied schoolmates with their lunches of store-bought bread. That must have been my mother's experience as well, something she didn't want me to share. Years later, when I asked my students in New Mexico to recount their lunchbox experiences, I heard similar stories, this time about tortillas. Hispanic students remembered eating burritos (tortillas of corn or wheat wrapped around meat and vegetables) while their Anglo schoolmates ate bread. Their experiences replicated the experiences of my mother and other children with sandwiches of homemade bread. Not by bread alone, but sometimes especially by bread, were the burdens and privileges of class and culture transmitted to young children.

Seven other authors have joined us in creating this collection, which combines women's history with the field of food studies. A series of national and international conferences brought us together in a common effort to document the roles of rural women. By sharing the results of our independent research, we found that during the late nineteenth and early twentieth century a group of professional women had emerged as experts devoted to improving the lives of rural women and achieving recognition for their work. They hoped to help women combine the business of producing food with their tasks as mothers and homemakers.

The authors of most of the essays focus on the work of individual women who came to prominence through their professional work. They situate these women in the rural cultures out of which they emerged to become experts. Our common task was to describe and understand how these late nineteenth- and early twentieth-century professionals viewed the lives of rural women. That process also gave us an opportunity to reflect on how we, as today's scholars and professionals, could carry on the work of those rural women who created the cultures to sustain the production of food.

It has been a pleasure to work with each of the authors, and we would like to thank them for agreeing to include their work in this collection. The history of food is sometimes complicated, and explaining the roles that women have played in that history is an important part of the story. We appreciate the support we received from our editors at the University of Iowa Press. From the first time we corresponded with Catherine Cocks, she saw the potential in our work and she has been a champion of this project. Her thoughtful suggestions, along with those of the anonymous reviewers, helped us to delve more deeply into the themes of professionalization and expertise among rural women. We also enjoyed working with Susan Hill Newton, our managing editor, and Carolyn Brown, whose careful copyediting made this book better. Catherine Plear, our indexer, is delightful to work with, not only because she embraced this project with enthusiasm, but because she loves ideas and good conversation. We gratefully acknowledge funding from the Laurentian University Research Fund that supported the last stages of this publication. *Women in Agriculture* is dedicated to our colleagues in the Rural Women's Studies Association because their scholarly work and active engagement highlights the women who occupy central roles in rural economies and communities—both past and present.

LINDA M. AMBROSE, Sudbury, Ontario
JOAN M. JENSEN, Las Cruces, New Mexico
Summer 2016

Women in Agriculture

Introduction

LINDA M. AMBROSE AND JOAN M. JENSEN

Women have always been crucial actors in the production, distribution, and consumption of food. Despite regional and national differences, rural and small town populations were still the majority during the first half of the twentieth century, the era we focus on in this collection. Women—most of them from rural backgrounds themselves—have also claimed a place as educators, commodity specialists, agricultural representatives, and leaders of voluntary organizations. The essays in this collection explore these aspects of women's history and its relation to the growing field of food studies. Previous studies have focused mainly on women as homemakers purchasing and preparing food, as mothers concerned about nutrition for their families, and as cultural agents who create and preserve particular foodways.[1] *Women in Agriculture* focuses instead on the largely untold history of women as agricultural professionals. The women we study provide examples of work in areas that are central to food studies, including social research, the science of production, and community engagement. Thus, this book will contribute to a fuller understanding of the emerging professions and disciplines relating to food during the period from 1880 to 1965.

Early agrarian professional women worked with food in commodity science, rural sociology, and marketing of agricultural products. Others were social scientists who surveyed and interviewed rural women about their roles as farmworkers, producers, and marketers of food products. Education, both formal and informal, was crucial to changing attitudes about what were considered appropriate kinds of agricultural work for women. From the Langham Place feminists of the nineteenth century to the promoters of female-run markets to community activists, agrarian professionals played key roles in suggesting new possibilities for women's engagement with food. Harriette Cushman, Emily Hoag, and Lorian Jefferson represent women who helped develop various agricultural sciences. Home demonstration agents, instructors of short courses, and organizers of Women's Institutes converted the recent research into forms that could be understood by lay audiences, primarily women. Most of these professionals believed in the power of women's collective efforts to transform their own lives and reform their communities. Rural women as producers and marketers of food needed support for these tasks as well as for their roles as homemakers and consumers, tasks they shared with the growing population of urban women. The information that rural women received from educators, organizers, and peers served to enrich their own experiences as farmers, homemakers, and community organizers. The information rural women shared with professionals also helped shape marketing policy. Organizations for rural women were key, not just for information transfer about food preparation or production schemes, but also for the collective self-confidence they engendered when local women strategized about how to improve their own conditions. Although rural women socialized with and learned from extension agents and home economists, they applied that knowledge in surprising and often unforeseen ways.

These essays on agrarian women, while an important addition to rural and food studies, also expand the concept of the New Woman, which has largely been conceived of as a strictly urban phenomenon. These New Women took up a variety of issues, such as their right to college education, engagement in "public" roles as opposed to "home" work, better pay and working conditions, new attitudes toward marriage, and wider acceptance for women's participation in nontraditional physical activities, including recreational activities and physical work. Scholars have explained that these New Women, who carved out career niches within both governmental and other public institutions, were usually young and unmarried. They claimed the right to a place in the middle-class occupational hierarchy previously reserved for white males. The first generation often worked in cities, yet many of these activists had deep roots in rural and small-town communities,[2] and, as we will show, some remained committed to working with rural women.

Although scholarship on the urban New Women provided key ways to look at professionals who worked in cities between 1880 and the 1920s, it ignored rural professionals. The women who engaged in rural professional work shared with urban New Women a defiance of conventional gendered norms by entering careers previously closed to women. *Women in Agriculture* looks beyond urban reformers to those who spoke to the still rural majority of the population and played key roles in professions central to food studies.[3]

Rural women's history is a field that has experienced significant advances over the past forty years. As with most other genres of women's history in the 1970s, researchers had first to overcome the common argument that sources did not exist to write histories of rural women. An early volume that refuted the argument that sources did not exist was Joan M. Jensen's book, *With These Hands: Women Working on the Land*, published by Feminist Press in 1981. Jensen organized more than 250 excerpts from writings by and about rural women, proving that a plethora of sources did exist. Since then dozens of historians have creatively explored the archives that already existed, created new archives through oral history, and crossed disciplinary borders to employ a variety of methods to explore in more detail the various roles rural women occupied as producers, partners, and protesters. *Women in Agriculture* continues that exploration.

Over the past almost four decades of research, scholars have laid the groundwork for rural women's studies, pointed the direction for further research, and suggested ways to incorporate interdisciplinary methods. Following the publication of *With These Hands*, a conference on rural women's history was organized in 1982, bringing together almost 125 scholars and rural women. From that beginning, new questions about women's role in rural life helped to expand the focus from rural women's history to rural women's studies. Conferences continued every few years during the 1980s and 1990s, culminating in the formation of the Rural Women's Studies Association (RWSA) in 1997. Since then, triennial conferences have been held in different regions of the United States and, in 2012, in Fredericton, New Brunswick, Canada. The next RWSA conference is scheduled for 2018 in Athens, Ohio. Although the largest contingents come from the United States, the conferences always include participants from Canada, Europe, and often Asia and Africa. Meanwhile, the RWSA has also sponsored and encouraged sessions on gender and rural women at the annual Agricultural History Society conferences. Other collaborations have included panels at the Berkshire Conference on Women's History in Toronto in 2014 and at meetings of the European Rural History Organization.

Our present collection reflects those transnational conversations, which have expanded the history of rural women from the United States to other countries, and continues them by uncovering previously untold stories and contributing to discussions and debates about feminism in rural settings. It advances our understanding of female experts, women's collective action, and the local responses to advice offered from state and educational authorities. We focus as well on rural women's greater participation in postsecondary education, paid work, and public roles. The essays in this volume profile women whose work was embedded in specific national contexts so as to form a collective biography of women who graduated into a world that was not always prepared to welcome them into the public life that professions demanded. It was also a time when various academic social sciences—economics, sociology, and political science—were emerging. Middle-class men were already creating these new disciplines and prescribing more traditional gender roles for these New Women. Professional women contributing to food sciences, commodity production, and community outreach sometimes encountered opposition from men (a resistance we call the "new patriarchy"). At times, however, these women received important assistance from men, especially those who shared a common rural background and an interest in rural life and agricultural production. Given the complexity of this history of women entering rural professions related to food, it is important to explore both practice and policy through a lens that is gendered.

Thus, a primary goal of our book is to emphasize the intersection of food studies and gender studies. The scholarship of the contributors forms part of the ongoing conversations within various disciplines of history—agriculture, gender, education, and public policy. By joining these ongoing scholarly discussions to food studies, we introduce new issues not always recognized as crucial to food studies, in particular the work of various rural professionals who made major contributions to food production, food security, and food science. The essays recover untold stories of women significant to history in various ways, but most important, the collection emphasizes how food studies can be enriched by paying particular attention to gender.

We hope that both scholarly and more popular audiences will find intriguing this group of women who performed in various intellectual and professional sectors. Our collection includes United States, Canadian, and Western European case studies of these previously neglected professional women who worked in rural settings. Mostly middle-class and Caucasian—with some important exceptions, such as African American and Native American women in the United States—the women we study were part of the growing transatlantic partnerships that also fostered both domestic and foreign colonialism. Despite the national differences, there was a dovetailing of economic and social thinking on both sides of the Atlantic, especially evident during times of war. As the New Women we study found ways to interact with their male counterparts, many of whom did not welcome women's presence, sometimes they shared racist, colonialist attitudes, but their own position as an unwanted minority also gave them a different view of the developing modern economies and societies. Indeed, agrarian professional women insisted that rural sociologists and other social scientists needed to hear the voices of the rural women in local communities where they worked.

This volume seeks to untangle some of the complicated ties between women and food. Because they were constantly looking for ways to increase their incomes and economize on their household expenses, rural women's relationships with food were complicated. As these essays reveal, the roles of women as producers and consumers were not distinct and discrete. While the rural producers sought strategies to get their best-quality products to market so that they could command the best price, those same women were consumers of food as well. There was a chain of interdependency that tied producers and consumers together in particular ways. Black women in rural Arkansas, for example, made efforts to expand their own production beyond cotton to include foodstuffs for their families and for market.

When scholars explore the ways in which experts interact with rural people, it is easy to make the assumption that the exchange is a classic case of social control, where experts impose their knowledge and their ways onto local rural women. Examples to the contrary, however, abound in this collection. Black demonstration agents who worked for the U.S. Department of Agriculture (USDA) gave instruction and demonstrated preferred methods, but Arkansan women modified instructions to suit their local circumstances. The case was similar in the Netherlands, where innovations were presented to rural women, but because of particular economic circumstances and family dynamics, women either ignored the suggestions or only managed to introduce change incrementally given the constraints of family dynamics. Often they had to please a husband's palate or comply with a mother or mother-in-law who continued to preside over the domestic space, thus slowing the pace of change. Instructors observed this incremental implementation and recognized that their rural students were working within a set of complex social realities. Perhaps one of the most striking examples in this collection of respecting local circumstances and acknowledging local ways of knowing is the case study in chapter 10 of Densmore and Warren, who worked together to produce an important book about indigenous uses of plants. Densmore depended on Warren's expertise as an indigenous woman to instruct her on botany by disclosing the practical uses of particular plants. Together, they wrote down the knowledge that Warren communicated and as a result created an important text.

Also complicated are the relationships that rural women had with experts. Indeed, understanding the question of "expertise" among rural women is itself a complicated matter. National bureaucracies gave rural women access to experts by offering instruction through a variety of state-financed initiatives in the United States, Canada, the Netherlands, and Great Britain. But the process of delivering information to rural women was never completely a one-way exchange. Some individuals were tagged as "experts" either by virtue of their employment as educators at agricultural schools or land-grant colleges, as commodity specialists or policy advisors, or as home economists named by the USDA extension service, the provincial department of agriculture, or the national agricultural service. Others drew their expertise from a variety of sources. Mabel Webb was able to establish her cultural currency from her war work among rural women, and she channeled that into a successful broadcasting career in which she shared the culinary expertise of other women. Adelaide Hoodless wrote a textbook for Ontario schools after her own child died because of improper food-handling practices in her home. Madge Watt declared herself a "rural expert" because she had several strategies in mind for empowering Canadian and British rural women to make money from their horticultural and agricultural efforts. Densmore collaborated with an Ojibwe elder to collect information on traditional uses of plant food.

None of those efforts to collaborate were problem free. Webb found her folksy household talks on radio challenged by some male scientists of the day, such as V. H. Mottram, who criticized her hints about economizing by reheating meats or stretching the food budget with peer-based advice. The Women's Institutes, transplanted from Canada to the United Kingdom, have often been gently mocked for their ubiquitous tea parties and cakes, but Andrews has argued that the movement became "the acceptable face of feminism." Although naysayers have sought to trivialize it, the movement's recent centenary celebrations in London included her Majesty Queen Elizabeth II, who acknowledged the many accomplishments of the group for village life, social issues, and national concerns. When elders and others questioned Densmore and Warren's collaboration at the time of their project, they pursued it anyway. The text they created recorded indigenous knowledge and made it accessible in a way that remains unmatched. The authors could not have foreseen the threat that biopiracy would pose in the coming decades.

Home demonstration agents and home economists are central to any study of rural women and the professions. Scholars such as Carolyn M. Goldstein have taken the study of home economics quite far by retrieving it from the dustbin where some mainstream feminist scholarship had discarded it.[4] While Goldstein's work concentrated on the period since the 1920s, many of the papers in this collection deal with the period before that era. Emily Hoag's work, for example, illustrates that before the rise and dominance of home economics, farm women's experiences were regarded as more complex and complicated than simply running a rural home under less-than-ideal circumstances. Yet the USDA's decision not to publish Hoag's findings meant that her study was relegated to the archives for several decades. Only when scholars discovered that treasure trove of data about rural women did Hoag's study begin to inform a different picture of rural women's experiences.

The essays in this volume point to other surprising discoveries as well. It seemed progressive, for example, that the work of home demonstration agents in the American South should be desegregated after the 1964 ruling in the United States that mandated it. Jones-Branch points out, however, that this seemingly progressive move had surprisingly negative consequences in some cases. McKinney's work on Harriette Cushman concludes that although she worked long and hard to establish standards for egg production and marketing in Montana, in the end, that work did not have long-term consequences. While Densmore and Warren's work at recording indigenous knowledge was laudable, the danger of biopiracy remains very real.

Food security is a theme that is woven through many of the essays because, from the beginning, women's role in ensuring food provision was crucial. Women acquired most of the food needed to make their families and communities secure. They did so either through producing it on the land themselves or purchasing it or some combination of the two. During particular periods of history, providing food security took priority over quality, particularly during wartime. But even during periods of relative peace, women's role in producing and preserving was central. This was as true in the Netherlands as it was in England, Canada, and the United States. Helpful advice about how to stretch the food budget came in many forms. It included tips passed between peers at Women's Institute gatherings, short courses on canning and preserving taught through the USDA extension service, Dutch state programs, or English radio broadcasts. Resourceful experts instructed women on how to make multiple meals from the meat they had on hand and to preserve the bumper crops in their own yards and village greens. Production and sales of surplus food could ensure family survival in rural areas, and sometimes women could generate substantial incomes through their food production. In Montana, the humble hen could quietly and dependably pay the grocery bill for items that farm households could not produce, while in British Columbia bees and cows helped women ensure a reliable cash flow. Across North America and Europe, female farmers not only economized, but they could—and often did—generate respectable incomes.

Women did more than just work to produce food. Sometimes food provided a welcome opportunity for sociability, cultural celebrations, and fun. Members of women's groups looked forward to a monthly cup of tea with neighbor women, and producers welcomed the opportunity to attend state fairs and commodity shows. Short courses offered not only expertise in food production but opportunities for families, kin, and neighbors to gather and strengthen community bonds. Everywhere, families and friends gathered for annual festivities and celebrations. As Linda Ambrose argues in "Forever Lunching," women gathered not only to eat and to chat but to discuss serious and difficult topics such as political rights, farm finances, segregation, war losses, community infrastructure, and the challenges of living on the land. When they met over the teacups, in times of peace and war, women learned from each other as well as from the experts.

Women had always gathered, planted, reaped, cooked, and shared food with their families and communities. New in the second half of the nineteenth century was an international market in food that, when combined with natural disasters of drought and pestilence, intermittently interrupted the traditional distribution of food. Women in the developed nations—the United States, Canada, and most of Europe—did not have to face famine during those years. Except for a famine in Finland and northern Sweden in the 1860s and one in Alaska in the late 1870s, there were no extreme shortages of food. In nations controlled by Western colonial powers, however, food security was often threatened. In India, some six million people died of starvation between 1896 and 1902. Chinese women faced extreme disruptions of food supplies because of war and drought during the first decade of the twentieth century. There were no major international wars in this period, however, and with a few important exceptions, women worldwide experienced relative food security. World War I interrupted food distribution in Lebanon, thus causing starvation and wartime food dislocations in Russia, which were increased by domestic unrest following the war. Some of the first international efforts to relieve food shortages emerged following World War I, a topic we do not cover but which was the foundation of an important international movement to deal with food insecurity.

In many developed countries agricultural communities changed as market economies grew and farm families adjusted to more commercialized agricultural systems. Yet important regional differences remained, and women of each region and nation faced new conditions. The United States provides one picture of those diverse conditions. In the South, black women farmers, especially, faced deteriorating racial relations and had greater difficulty than white women getting and retaining land. Many women turned to truck farming after the Civil War, raising small animals and vegetables to market locally or selling baked products. Reports from the South indicated that some black women canned fruit and found a ready market for their processed food. By the 1890s, however, getting and controlling land was increasingly difficult, and many were desperately looking for alternatives to the lives they faced as sharecroppers tied to a system that gave them little room to succeed as farm owners. Should they head north? Try to obtain land in the West and create black communities that they could run themselves? Or, head to Africa, where some felt they would have a chance for freedom?[5]

White women farmers in all areas of the United States faced new challenges as well. In the Midwest, settlers who took land from indigenous people faced continuing challenges as they moved west. Homesteaders found soil that needed deep plowing with machinery and sturdy workhorses. To send crops east by train took additional money. As soon as one area was settled, land prices rose, making it hard to provide land for children even as the number of children in farm families declined. In the Northeast, women tried new crops, established truck farms, raised poultry, and planted orchards. Women everywhere turned to improving their traditional skills at dairying, producing butter for local markets, and raising poultry. They worked together to improve their rural communities and sent more of their children to school for a longer time when they could. Still, their political fortunes seemed to be more controlled by industrialists and the growing urban areas. By the 1890s protest groups—the Grange, Farmer's Alliance, Populists, and the Country Life Movement—were forming in rural areas. Women belonged to all these groups and participated in protests and new farming experiments.[6]

To demonstrate the wide variety of professions that women took up related to food, *Women in Agriculture* is organized around three themes. Part 1, "Education," focuses on women's fight for access to pragmatic education about farming and to careers as educators, administrators, and researchers in institutions of higher learning and state-sponsored programs. Part 2, "Experts," offers case studies of women who used their education to create nontraditional roles for women in agriculture as commodity experts, rural sociologists, and policy advisers. Part 3, "Extension," turns to the important work of outreach, in which women worked as broadcasters, agricultural agents in state programs designed for rural women and men, or directly led community organizations, often with attention to food security, viable rural economies, and community organization.

Part 1: Education

According to Karen Sayer and Nicola Verdon, in England calls for women to enter the farming profession were rooted in the late nineteenth century. Arising from a feminist commitment to expanded horizons for women in all arenas, the women of the Langham Group argued that rather than drudgery or—worse yet—immorality, the work of farming could be a respectable undertaking for women with adequate background experience and professional training. The business of farming, they argued, not just field work or the more "feminine" pursuits of dairying and beekeeping, could be respectable and lucrative enterprises for women. What was needed was adequate training and education to allow women to compete with men on equal footing in agricultural production. Sayer and Verdon make a persuasive case that the arguments presented in the pages of the *Englishwomen's Review* laid the groundwork for later developments in agricultural education.

Arguing that women played important roles in the fields of rural sociology and agricultural economics, Joan Jensen focuses on the example of Emily Hoag at the University of Wisconsin at Madison. Hoag participated in research that guided food policy in the United States, and her work helped determine women's place in food production and food policy. Jensen shows that Hoag's work, both as a university instructor and as a research assistant, made her a key contributor to the development of the discipline of rural sociology. Although Hoag arrived in Madison as an experienced classroom teacher with a family farming background ideally suited to carry out research among farm families, the professors who published the results of her fieldwork usually recognized her only in a footnote. When the Wisconsin professors, all men, created their own history of agricultural economics and rural sociology, Hoag was seldom mentioned. By tracing Hoag's work, Jensen offers insight into the previously overlooked roles of New Women in university research settings.

Margreet van der Burg continues this attention to education. She explains how both male and female agricultural reformers in the Netherlands succeeded in establishing farm household management education with the national vocational system. By implementing curriculum suitable for different types of farms and social classes, the state and agricultural organizations acknowledged that farm women and other rural women were not homogenous. The qualified female teachers of the same background became their leaders. This way, many women participated in a wide variety of initiatives that challenged notions about women and food by emphasizing that women were not merely consumers or stereotypical "dairy maids" but central actors in the production of food stuffs and thus crucial participants in the rural economies.

Part 2: Experts

In the second part of the book, authors focus on individual women who played central roles in the history of food in North America. Amy McKinney opens this section by exploring the career of a commodity specialist, Harriette Cushman, who made efforts to develop marketing and distribution strategies for Montana eggs from 1922 to 1955. As a poultry specialist, Cushman worked with poultry women, encouraging them to standardize the eggs they marketed by incorporating scientific candling and grading procedures. Cushman fostered new legislation to establish quality standards to guide the producers of this food staple. Women continued to dominate egg production and marketing for most of the early twentieth century.

Anne Moore's chapter introduces Lorian Jefferson, who joined the faculty of the Massachusetts Agricultural College in 1912, where she conducted research for more than twenty-five years. Moore argues that Jefferson's dedication to scientific and empirical investigation of New England food production, supply, and distribution offers an important example of a "new agrarian woman." Through her research and publications, Jefferson not only represented a woman occupying a nontraditional role but also one who "paid close attention to women's roles as producers and consumers in a new era of food availability and consumerism." Indeed, Jefferson and Emily Hoag had a great deal in common, including their connections to Madison, Wisconsin, and their ties to the Country Life Movement.

Linda Ambrose then looks at three Canadian women who were leaders in the Women's Institute (WI) movement: Adelaide Hoodless, Laura Rose, and Madge Robertson Watt. Ambrose reminds us that the WI, along with other groups for rural women, took up at least three different issues of interest to food studies—food science, food security, and food production for profit. All three women gained reputations as experts through their lectures, publications, and efforts to organize rural women, yet they followed very unconventional career paths. The learning they applied to rural situations arose in large part from their own life experiences. Adelaide Hoodless became an expert on domestic science because of a personal tragedy connected to the improper handling of milk. Laura Rose gained her expertise at dairy work through her time on a family farm in the United States; and Madge Watt, the Canadian founder of the WI in Britain, declared herself to be an expert on rural matters because her university education and journalism career had trained her to look for every available opportunity for women to broaden their horizons. These Canadian women set the stage for later transnational work, including founding the WI in Britain, where the movement led to a series of retail markets run by women.

Part 3: Extension

One of the most common goals the experts shared was a commitment to extend educational opportunities to rural women in their own communities through demonstrations, short courses, and later by broadcasts designed specifically for rural audiences. Often, women professionals who were involved in the distribution of food knowledge simply called it "extension." In turn, however, they learned much from rural women, including those indigenous to the land. The women who taught such courses, whether at colleges, in communities, or on the airwaves, were often the women we discuss in part 2 as "experts."

As Maggie Andrews explains, Mabel Webb was not formally trained in the culinary arts, but she was a self-proclaimed expert on cooking because she traveled far and wide across England enjoying regional foods and gathering recipes, both tasty and economical, to share with her radio audience. In discussing the role of "Mrs. Webb," who became popular in Britain after World War I, Andrews explains how this radio personality empowered women through her folksy radio talks about recipes and food. Although authorities criticized Webb for lack of scientific method or formal domestic science models, she wielded an enormous influence over her listening audience—a popularity that perplexed the executives at the British Broadcasting Corporation, who considered Webb's assertive style to be overly "aggressive." A New Woman in her own right, Webb commanded attention from both her loyal listeners and those who tried to confine or redirect her voice.

Cherisse Jones-Branch then describes how African American home demonstration agents in Arkansas created their own space for expanding traditional "extension" classes. From 1914 to 1965, these agents instructed women about food, its production, preservation, and preparation, as the Agricultural Extension Service expected them to do. But as Jones-Branch explains, in these segregated groups, agents also incorporated the concerns of local women that went far beyond these common concerns. Some scholars have argued that extension agents were a part of the long arm of segregation as they worked in the South. Jones-Branch argues that these extension programs under black demonstration agents also offered women opportunities for self-improvement. The roles and relationships established by black demonstration agents were complicated, and Jones-Branch provides an example of how agents hired to teach traditional female tasks, such as canning food, might go on to give instruction in other health-related issues, including making birth control information accessible to rural women.

The impact of local women meeting together to receive instruction from extension agents or other food experts in seemingly traditional womanly arts should not be underestimated. Food was almost always at the center of such gatherings, and outsiders sometimes dismissed such gatherings as simple idle pursuits, deemed to be especially inappropriate during wartime. Linda Ambrose demonstrates that groups such as the Women's Institutes in Ontario could also function as a front for something far more political than they seemed. Serving food at meetings, particularly the elaborate rituals of tea pouring and cake making, might appear to be only a frivolous cultural practice. In fact, the consumption of food in these community settings provided a cultural space for women to exchange ideas about a variety of issues, and the funds they raised from selling food presented opportunities to strategize about various community-based initiatives that the women themselves controlled.

Some rural women, however, instructed the experts in a type of reverse extension. Joan Jensen discusses the instruction about food plants in indigenous cultures through the collaboration of Mary Warren English, an Ojibwe, with the nonnative anthropologist Frances Densmore. Densmore is known as the author of *The Uses of Plants by the Chippewa Indians*, but less is known about Mary Warren English, who worked with her collecting traditional knowledge about the plants and medicines that formed the basis for the book. Although English faced some opposition from indigenous people about recording traditional Ojibwe plant knowledge, she was convinced that writing it down was an important way to preserve that knowledge. Densmore and English's partnership suggests that the transfer of knowledge was sometimes the result of collaboration by women who felt that documenting native ways of knowing about the nutritional and healing properties of particular plants was an important way to preserve that knowledge. The work of Densmore and English foreshadowed the ongoing battles that rage between pharmaceutical companies, who are eager to capitalize on indigenous knowledge, and Native American communities, who claim sovereignty over that knowledge.

We hope that readers will enjoy this collection of essays focusing on women and food. It should broaden the understanding of how women historically have been intimately involved with food not only as producers, preparers, and consumers but also as food specialists, rural sociologists, policy analysts, demonstration agents, broadcasters, and community activists.

Part 1

EDUCATION

1

The Professionalization of Farming for Women in Late Victorian Britain

The Role and Legacy of the Langham Place Feminists

KAREN SAYER AND NICOLA VERDON

Women producers were, as Ivy Pinchbeck argued in her 1930 book *Women Workers and the Industrial Revolution,* at the forefront of British innovations in agricultural food production.[1] In Britain, as in the rest of Europe, the United States, and Canada, subsequent wide-ranging secondary literature has addressed the story of women's participation in agricultural production. These critical and nuanced assessments of the varied and changing position of women have, for the most part, focused on the paid work of women employed in the fields.[2]

This is perhaps not surprising. Despite women's involvement in dairy work and their being encouraged to join other areas of food production, such as fruit growing or honey production, for most of the modern period, when women's participation in farming came to the attention of the British public, it did so primarily with reference to laboring women who worked in field and dairy for a cash wage. Women's employment in agriculture was, for example, surveyed by Parliament in 1843, an exercise that was repeated during the late 1860s. The findings of the parliamentary reports were disseminated in the local and in the national press and by social commentators, who added to the available body of evidence through their own independent research. In most of these reports and publications, the impact on women of their labor was the issue at hand, and field labor (the most visible employment of all) was attacked as unsexing and morally corrupting. In 1867 Charles Dickens wrote that field work "converts girls into demons." He was far from unusual in this assessment.[3] As a result of this kind of rhetoric, specific forms of paid agricultural labor were restricted by the Agricultural Gangs Act of 1867 (30 and 31 Vict. c. 130).[4] For the most part, the only dissenting voices in these debates, defending laboring women's right to work, were those of some farmers, a few independent observers, such as Arthur Munby, and the campaigners of Langham Place.[5] This chapter focuses on these campaigners, who came together at Langham Place in London in the late 1850s and who, by the 1870s, were making a unique case for well-to-do women joining the farming community as professionals.

The Langham Place group, a small circle of middle-class women who were linked into the literate, educated elite circles of their day, converged around the leading figures of artist and campaigner Barbara Leigh Smith Bodichon (1827–1891) and her friend, the journalist and activist Bessie Rayner Parkes (1829–1925), who described Bodichon as "the most powerful woman I have ever known."[6] Prefiguring the interests, concerns, and campaigns of those who would be characterized as examples of the New Woman, in 1858 the Langham Place feminists published the *Englishwoman's Journal* to disseminate and discuss the groups' shared interests in widening middle-class women's spheres of work, education, and training. The Langham Place women were largely from metropolitan, bourgeois, and broadly Liberal and radical backgrounds (Parkes, for example, attended a Unitarian boarding school), but from 1859 they were joined by one not cut from this mold: Jessie Boucherett (1825–1905), the daughter of a Lincolnshire landowner and a strong Conservative, who rode to hounds boldly and inherited her father's six-thousand–acre estate upon his death in 1883.[7] Her involvement was pivotal. Along with the poet Adelaide Proctor (1825–1864), Boucherett ran the Society for the Promotion of Employment for Women from the Langham Place offices.[8] In 1864, when the *Englishwoman's Journal* ran into financial troubles, Boucherett inherited and edited the publication from 1866 to 1871, renaming it the *Englishwoman's Review* (publication continued until 1903).[9] Employment had always been a major theme of the journal, but it was during Boucherett's editorship that debates around food, farming, and agricultural production started to gain real prominence. Within these exchanges a number of issues that concerned the Langham Place group more generally were revealed: the right of single women from all classes to earn a living, the extent to which married women's employment should be encouraged, the appropriate division of labor between men and women, and the impact that widening spheres of employment had on middle-class women's social status. Whenever farming and agricultural production were linked to food, production became noticeably more "respectable" for women. The *Journal* even published an article by African American reformer Frances Harper recounting the agricultural successes of recently freed former slaves in the American South.

The Langham Place group was far from being the first to outline the advantages of farming as a form of food production well within the woman's domain. As K. R. and G. E. Fussell note in their account of rural women's lives, the *English Countrywoman* (1953), the anonymous author of *Farming for Ladies* (1844) suggested that women should keep a cow, pigs, and poultry in order to enjoy the benefits of the best cream, butter, and eggs. The author counseled the "mistress" not to entrust their care "wholly to domestics" but to "rise early to superintend the feeding of her chickens, the milking of her cow, the churning in her dairy."[10] This treatise was typical of its kind in that it was aimed at relatively wealthy women more interested in finding a diversion than in the practicalities of running a commercial farm. The Langham Place group's approach was quite different. While building on the traditions of elite women's management of the dairy and poultry yard, they increasingly saw agricultural production on a wider scale as an equally suitable and lucrative business for the well-to-do woman. As a result, farming was represented in the *Englishwoman's Review* as an enterprise fit for women and as a profession requiring education, skill, and training. This claim was achieved in part through the association of farming with (apparently) smaller-scale forms of agriculture—dairy, farmhouse, orchard, and garden—that were tied to (respectable, feminine) food production but that also leapt over or remapped the boundaries of these smaller-scale spaces. Building on an understanding of all women's employment as grounded on rational, economic decision making, coupled with a capacity for skilled work, the Langham Place group perceived middle-class along with working-class women as capable of both highly skilled and laborious productive work in all spheres. Therefore, they represented women as capable and intelligent persons who could not only grasp the latest technological advances in farming but also write and lecture on them and win prizes for their application. They were, however, careful to make a distinction between field labor, undertaken out of economic necessity by working-class women, and farming, undertaken as a professional pursuit by educated women. The women of

Langham Place argued that despite the complexities of farming, respectable women in nineteenth-century Britain were perfectly suited for it.[11] This position went against the convictions expressed at the time by many male commentators, who argued unequivocally that women involved in agricultural pursuits ought to confine their efforts to the smaller-scale, more feminine sphere of the "subsidiary" branches, such as the garden, apiary, poultry, or dairy production, because they were physically and mentally unsuited to and incapable of managing large arable enterprises or livestock farms.

By making their case, the Langham Place circle aimed, as always, to expand the arenas seen to be suitable for women but couched their suggestions in terms of domesticity and home so as to be less threatening to male authorities. They reinforced the case for expanding women's involvement in the business of agricultural production—and also the wider case for agricultural production's standing as a profession—by using emerging ideas about rational and scientific approaches to food production, physical well-being, and family health. When women like Jessie Boucherett and Barbara Bodichon defended women's right to work within agricultural production and argued that women ought to take advantage of the agricultural depression by becoming farmers in their own right, they thereby produced new maps of professional identity that refigured the accepted, highly gendered, relations within food production and the relationship of (food) producer to consumer. This remapping laid the foundation for later campaigns for women's education and the right to a professional standing within specialist fields of food production, including those fields traditionally thought of as "feminine," such as dairy and poultry production, which at the time were treated dismissively in official reports.[12] The Langham Place argument, by engaging with farming as a profession, also contributed to the wider debate about the standing of agriculture as a whole as a profession (not just a trade, industry, or tradition). Before making the case for women as producers, however, they initially entered the debate at the level of women's employment on farms.

Defending the Right of Laboring Women to Work

In Britain, food production—in the form of agriculture, grazing, dairy production, fruit, and horticulture—came to prominence as an occupation of national significance in the late eighteenth and early nineteenth centuries. The campaigns and authorship of leading agricultural commentators such as Arthur Young (1741–1820)[13] and William Marshall (1745–1818),[14] political-economic texts such as Harriet Martineau's *Brooke and Brooke Farm* (1833),[15] the formation of the Royal Highland and Agricultural Society (established 1784) and the Royal Agricultural Society of England (established 1840), and their equivalent county associations[16] were all significant in writing and reflecting on what subsequently came to be called the Agricultural Revolution. These authors accepted the sexual division of labor in agriculture as a normal and an efficient component of the system until at least the 1830s and 1840s. The work that women undertook varied by region, season, and over time, but the division of agricultural labor was a constant, even as new technologies and new methods of farming were introduced. Among farmers, most were tenants who occupied land rented from landowners or the Anglican Church. Many employed family labor, but also farm servants (men and women who lived and worked on the farm), and male day laborers. Female farm laborers, who were employed on a casual daily basis for approximately half the wages that men received, were used in field work, especially during the hay and corn harvests. The only field women to have any surety in this class of employment were the "bondagers" of the Borders (of the North of England and Lowlands of Scotland). By the nineteenth century, "bondager" referred to a woman who was taken on by a "hind" (male agricultural laborer) at the annual hiring, to fulfill his contract to the farmer. The hind would have a cottage as part of his contract, and the bondager either lived with him and his family or in a cottage with other bondagers nearby. It was regionally specific hiring system grounded in the topography of the area that enabled farmers on scattered farmsteads to have a supply of regular labor throughout the year and access to additional labor at times of shortage, such as during the period of turnip hoeing in June and July and during the harvest period, though the practice seems to have been that work was found for them on the farm all year round. Northumberland was a place, according to the *Parliamentary Reports* on it, where education and high moral standards were supposedly valued, where male and female agricultural laborers were generally reckoned to be well-paid and materially well-supplied. Although the formal version of the contract eventually died out during the late nineteenth century, it was still common to refer to women field workers as "bondagers," and women continued to be employed in agriculture throughout the region into the twentieth century, in large part because of their excellent reputation for strength, reliability, and skill.[17]

When women's farmwork was first recognized as an issue by the *Review* in April 1867, the focus was not the bondager system of the North but another form of hiring casual labor that had gained a notorious reputation for exploitation and immorality: the East Anglian gangs. These "public" gangs were a form of subcontracting between a farmer who had a particular job to do (such as weeding corn or picking stones off the land) and a gangmaster who engaged a group of local women and children to perform the work for a daily rate of pay.[18] The *Review* piece was written in response to a parliamentary report on the subject that had just been released. The author accepts the official evidence that there was "disorder and immorality" among the laborers and that the "demoralisation [of field women] at present is great," but, unlike Dickens, she uses the dominant definition of femininity radically to praise the women, not to damn them. The article challenges the idea that women were meant to be physically weak and calls into question the representation of strong and tanned women as masculine and immoral. The author stresses that these women had to work, and within the context of the journal, which fought for the single woman's right to earn a wage without the interference of protective legislation, this was entirely acceptable.[19] "If deprived of this resource," the article continues, the condition of women "would be wretched indeed, as there is absolutely no other honest employment open to them."[20] As a result, the author denies that intervention is the answer and argues instead that a new law would only exacerbate the problem by increasing the number of men in agriculture and leaving women in a more vulnerable position.[21] The government did, however, enact a new law, the Agricultural Gangs Act of 1867.[22] This legislated in three main areas: it outlawed the employment of children under the age of eight, the employment of males and females in the same gang, and stipulated that no female was to be employed in a gang with a male gangmaster unless a licensed female gangmaster was also present. It was the only government legislation aimed specifically at curtailing the employment of agricultural workers in the nineteenth century, although it stopped short of prohibiting female and child work in gangs altogether, something that had been under discussion.

As Candida Ann Lacey has noted, the Langham Place group were initially mindful of the need, as they saw it, to maintain a certain level of respectability. Respectability would allow them both to argue reasonably for women's rights on a public platform, and to challenge the dominant constructions of femininity and gentility that prevented single middle-class women from taking paid work. Their solution was to provide access to new forms of respectable employment that could be made to fit within dominant concepts of femininity.[23] This tension often meant that the Langham Place group applied uncritically the same liberal noninterventionist criteria to working-class women's employment as they did to the important and continuing issue of middle-class women's "professions."

Rural working women reappeared in the journal in 1869. This new article centers on a discussion of the first report of the Commission on the Employment of Young Persons, Children and Women in Agriculture (the commission was set up in 1867, before the Agricultural Gangs Act). Initially, the article discusses the pros and cons of different hiring systems, approving of high wages, paid in kind, which prevented drunkenness, and the use of single women, who would then learn how to handle money and prevent a population increase (though they learned how to handle money, their earnings went solely into their families, so they never controlled it). The employment of married women does not, however, meet the author's approval. Finally, the article notes that women's work encourages men to emigrate by reducing their chances of employment in England. This would help the colonies and again slow the increase in population. Like the earlier piece, this article did not approve of married women's work because working wives could not cook good food for their family or educate their children. The answer was to pay the men more, so that their wives were not forced to go onto the land. That argument, like the support for emigration, coincided with the ideas of the National Agricultural Labourers' Union, founded in the early 1870s.[24] The author challenges the idea that women who work in agriculture are immoral. Statistics proved that field women were no worse than other working women, including domestic servants. Any immorality that did exist was to be blamed on society, not the female laborers themselves.[25] Further, the article argues that it was better to be doing this kind of honest work than to be forced into prostitution, a key issue of the day. "A certain immorality is said to result from every employment for women that can be named," the article claims, "but want of employment produces far more." It continues: "The occasional immorality of women workers in the field, in manufactories, in domestic service and elsewhere, is preferable to the wholesale immorality of those who have no work and consequently no honest means of earning their bread."[26]

Boucherett was one of the few women at Langham Place who might have known what work on the land entailed, although as the daughter of the High Sheriff of Lincolnshire, she was of the landowner rather than the working class. She became interested in the cause of women's employment after reading Harriet Martineau's "On Female Industry."[27] Boucherett gave evidence to the 1867 Commission on the potential of training girls (in domestic work) in Lincolnshire, and it is possible that she is the author of the 1869 article on rural working women.[28] A founding member of the Freedom of Labour Defence League, which fought protective legislation, this kind of noninterventionist approach and defense of the field of women's right to work matched her typical thinking on the importance to women of paid work grounded in both the respectability and the rationality of labor designed to produce a wage to buy food.

The *Englishwoman's Review*'s 1869 article concludes: "Any further interference would be an act of oppression towards the women."[29] This is a forceful statement of laissez faire values, centered on the right to paid work and recognition of the rational liberal individual as embodied by the working woman laboring to produce food. It deftly uses food to rhetorical effect. The author also uses the dominant concepts of political economy and self-help, notably the right to work, and the concept of physical and moral health of women and the population as a whole. Each concept, commonly applied in mainstream nineteenth-century political, social, and economic debates at this time, is turned to defend rural women working to produce food (honestly) both for themselves and their families and for the nation. The standard elements of dominant middle-class ideology are used to argue a case totally opposed to that ideology's definitions of femininity. The article also uses the working-class men's argument that women took men's jobs and wages to argue that reduced male employment would be a good thing. The subversion of these dominant concepts was radical and new within debates about agriculture.

Promoting Food Production as a Profession for Middle-Class Women

By the 1870s the *Englishwoman's Review* had turned from the issue of working-class women's agricultural labor (except for reprinting articles from other journals that supported women's right to work) toward the argument that women with some capital ought to take up farming in all its forms. As it warmed to the subject, the *Review* summarized international developments on the topic of women's training in all branches of agriculture, including dairy farming, cheese and butter making, fruit culture, and beekeeping. Overall, the journal promoted food production based on rational principles.[30] Langham Place publications on food production stressed the importance of specialist education and also at times the potential for women to be hired to teach in these emerging fields as a new profession in its own right. An entry in 1881 headed "Food Production" connected with the organization the Association for the Promotion of Food Production by Women, for instance, listed "lady teachers who are willing to take pupils and to board them" and included tutors in dairy and poultry keeping, beekeeping, and various aspects of gardening, around Britain.[31] But these articles, which belonged broadly to the journal's wider campaign to expand women's involvement in the professions and skilled trades and to increase their access to specialized education as whole, also fell into the context of, and contributed to, a much wider debate taking place within British agriculture at the time. This centered on the importance of agricultural education in the United Kingdom, the topics to be taught, examinations to be set within agricultural degrees and diplomas at the end of the nineteenth century, and the professional standing of the farming class as a whole.[32]

Farming in Britain, entangled though it was with occupation through tenancy, was not a new occupation for women. Barbara Leigh Smith Bodichon had delivered a paper, "Reasons for the Enfranchisement of Women," to the National Association for the Promotion of Social Science in 1866. During the course of the lecture, she read out a letter from Matilda Betham Edwards (1836–1919), arguing that wealthy women were prevented from investing their money in land because they did not have the vote. Landowners wanted their tenants to exert power in Parliament, which was impossible for the unenfranchised widow or daughter of a farmer:

> It is not perhaps sufficiently considered how large a proportion of women occupy and cultivate farms entirely on their own account, nor how sensibly a share in the suffrage would affect their interests. . . . Farming is a healthful, easy, and natural profession for women who have been brought up in agricultural counties, and have thus been learning it from childhood. Moreover, for holders of capital, it is a tolerably lucrative one. I know many and many a single woman living upon the narrow income derived from a fair property invested in funds, who would gladly hire land instead, and thus obtain a higher interest for her money. It seems to me not a little hard that a woman possessing capital should be deprived of the privileges other capitalists enjoy, but it seems harder still that she should be robbed of her livelihood, simply because an anomalous custom has shut her out from such a privilege.
>
> In conclusion, I beg to say that I have been a farmer for some years, that I know of few parishes in which women are not owners or occupiers of land, and that every practical farmer with whom I have discussed the subject of the extension of the franchise to women, has recognised the justice of such a claim. They certainly see no reason why we should be entrusted with property, and not entrusted with the influence pertaining to it.[33]

From Matilda Betham Edwards's letter and Bodichon's paper, one can see that some of those at Langham Place had already decided that farming *was* a suitable employment for women by the time that the legislation on women's agricultural labor was being debated. However, as Betham Edwards's letter makes clear, at this point it was only viewed as an appropriate occupation if the woman concerned had been brought up to it. All the examples provided in her letter were of widows and daughters who like herself had been left tenancies on farms that they had worked or lived on.[34] In fact, in part as a result of this practice, the number of women returned in the decennial census category "farmers and graziers" remained relatively constant throughout the second half of the nineteenth century. The 1851 census records that 226,515 farmers and graziers in England and Wales were male and 22,916 were female. Just under 10 percent of farmers were therefore women in mid-Victorian times. The number of women classified as farmers remained remarkably stable over the next fifty years, despite the fluctuating fortunes of farming during this era. In 1901, with 202,751 men and 21,548 women recorded as farmers, women still represented one in ten farmers. The vast majority of female farmers, like Betham Edwards, had been brought up in the family business, and most, perhaps three out of four, were widows living in households with their children. They were, as the 1901 census noted, "carrying on the farms formerly held by their husbands."[35]

The *Englishwoman's Review* began to address the specific issue of women's training to enter farming as a profession in the 1870s. Initially, the *Review* argued simply that widows of farmers should be allowed to occupy the farms that they had worked on all their lives rather than be left at the mercy of sons who might turn them out. Next, a farmer had written to the *Field*—a popular country and field sports magazine of the time—to complain that she was not allowed the benefits of joining the Royal Agricultural Society of England because she was a woman.[36] Middle-class farming women suffered discrimination in law and practice, and the *Englishwoman's Journal* was determined to support their cause. It began to list the prizes that women won for farming at agricultural shows, along with similar achievements, as encouragement and sisterly solidarity.[37]

Then, in 1879, Boucherett published "Agriculture as an Employment for Women" in the *Review*. The article was a move away from defending the rights of widows to stay on their farms to promoting farming as an occupation for suitably minded, capable women. She argued that the depressed state of agriculture at the time provided women a unique opportunity to take up farming because competition from established tenants was decreasing. If a landlord had three or four farms on his hands, she suggested, "the offer even of a woman . . . would have a good chance."[38] Boucherett partly picked up on the theme of Betham Edwards's letter, which she may have read, by suggesting that normally women could not get tenancies for two main reasons: first, they did not have the vote; second, landlords feared that they would be bad farmers. Boucherett felt that the first was now less of a problem because of the reduced number of farmers looking for property and that the second could be overcome if the woman had proof of her knowledge of farming and could provide a man to back her. She suggests that an interested woman should first of all find such a man and then try to get a farmer to give her resident tuition in the practice of agriculture over two years. At this point Boucherett emphasizes that farming is a skill in which one would fail through lack of experience and that it is an expensive proposition. But, she concludes, "an industrious and energetic woman, if possessed of country tastes, might lead a happy and useful life as a farmer."[39]

Boucherett's article sparked a public discussion with an interchange of letters, replies, and notes on the subject of farming and other country pursuits for women. Boucherett was careful not to underestimate "the difficulty of learning a new business." She emphasized the need to be able "to do the work" as well as to supervise it in order to run any successful farm; dairy farming would require at least a year to establish.[40] Farming was a physically taxing business pursuit which, along with the ability to run a business, was a characteristic usually represented as masculine in Britain at this time. Boucherett offered a clear challenge to any such gendered definition that might consign women to more traditionally feminine fields of food production grounded in the established sexual division of labor within agriculture. The Langham Place group saw all forms of food production as realistic, respectable occupations for those women who had money available. Further notes on women farmers and the promotion of women as farmers in Britain and Europe continued in the *Review*.[41]

In the 1870s the Langham Place feminists began to argue consistently for women with some capital to take up all forms of food production, including farming, and for their access to training in various branches of skilled agricultural work such as dairying. The Langham Place group worked on creating a new role for women: the female businesswoman and agricultural expert. They increasingly saw agriculture as a suitable and lucrative business for middle-class women. In the eyes of the Langham Place authors, grounded in their own experience, well-to-do women were capable businesswomen, who could learn the requisite techniques, grasp the latest technological advances in agricultural methods, write and lecture on them, and win prizes for their application. The Langham Place campaigners led the discussion about the suitability of agriculture as an occupation for middle-class women, and their writing brought women's involvement in farming and food production to the public's attention. Through their publication, the *Englishwomen's Review*, their concern shifted from a concentration on the working conditions of female laborers in agriculture to promoting farming as a suitable profession for comparatively genteel women, whose social backgrounds and physical environment stymied opportunities to undertake meaningful work.

The Legacy of Langham Place

The Langham Place group was part of and engaged with a much wider public debate about the nature and meaning of work as the "surplus woman question" emerged in Britain and suffrage campaigners such as Josephine Butler sought the vote to underpin professional qualifications.[42] They were unique, however, in the ways they showcased agriculture and food production as possible professions for women. It is nevertheless unlikely that the Langham Place debates mobilized women to take up food production in great numbers. The level of risk to both pocket and reputation in entering a business in which conventional gender norms were so entrenched was perhaps too high. Victorian farming was dominated by masculine language, imagery, and technology, and women who were involved in the industry as workers, wives, or managers were subject to much critical comment and even contempt, which the Langham Place group fought.

The significance of the Langham Place group lies instead in its being a forerunner of a later, more concerted campaign to educate, train, and place middle-class women in farming and food production. The movement reemerged in England a generation after Langham, in the 1890s and early 1900s.[43] This was a time when both the term "New Woman," coined in 1894 by British novelist Sarah Grand, and the figure of the New Woman quickly came to public prominence. The classic image (or stereotype) of the New Woman in Britain was that of an Oxbridge-educated, trouser-wearing, cigarette-smoking, bicycle-riding, emancipated young woman who sought to overturn traditional conventions surrounding women's place in society. As Gillian Sutherland's recent *In Search of the New Woman* has shown, however, the relationship between this caricature and actual social change was much more complex. The New Woman trope was tempered and reworked by different groups of women to fit a range of identities and agendas.[44] Those at the forefront of women's agricultural education were mindful of both the opportunities that food production offered to the New Woman and the assistance that the New Woman could extend to farming and rural life in general. Although there had been some advances in the fields of female employment and education, career opportunities for unmarried middle-class women were still incredibly narrow at the end of the century. At the same time, many branches of British agriculture were still struggling to emerge from the economic depression that had cast its shadow for more than twenty years, denting not only production levels and profits (particularly in traditional arable farming) but the rural population as well, including the farm labor force. Middle-class women from towns were therefore encouraged to embrace the "back to the land" movement, to take up food production, and in doing so, to infuse "new life and spirit" into English rural life.[45] The prospect of women taking up farming was seen as a potential boost to a declining national industry and an answer to the lack of remunerative and engaging work for women in urban areas. As one correspondent put it, "It is, I know, a bold thing to suggest that women of cultivated tastes should go 'back to the land,' though in these practical days I see no reason why we should be regarded as capable of merely earning money by needlework or by other essentially feminine occupations."[46]

Just as the Langham Place group had been careful to maintain a level of respectability within a discourse that challenged traditional gender constructs, campaigners at the turn of the century still had to tread a fine line between promoting farming as a suitable career for prosperous women and maintaining appropriate class and gender appearances. Indeed, in an era when the practicalities of agricultural training were being realized, some leading voices narrowed their expectations of what middle-class women could achieve. In an 1899 article "The New Women and the Old Acres," Lady Warwick (Frances Evelyn Greville, 1861–1938), argued that although trained and "cultured" women, could have a "salutary and energising" effect on agriculture as farmers, peripatetic lecturers, or teachers, they should not aspire to farming "large acres" but confine themselves instead to "the lighter branches of agriculture": flower and fruit-growing, jam-making, beekeeping, and poultry farming.[47] This was a theme taken up by others, including Edith Bradley and Bertha La Mothe, in their *The Lighter Branches of Agriculture* (1903).[48] Although this model of female participation was restrictive in some respects and still bounded by the traditional division of labor in agriculture by gender, campaigners recognized that "le petit culture" offered middle-class women a practical and potentially profitable route into farming. These branches of food production, once derided as feminine and subsidiary, had weathered the agriculture depression better than other types of traditional arable and livestock farming, and they began to expand as improved transportation networks made it possible to meet growing demand from urban markets. The expansion provided careers for middle-class women in work that was appropriate and fulfilling.

Unidentified students digging a celery trench at Studley Horticultural and Agricultural College, Warwickshire, UK. Studley was opened in 1903 by Lady (later Countess) Warwick. It was the successor to Lady Warwick Hostel, which had been attached to the University of Reading between 1898 and 1902. The college remained an all-women institution throughout its existence. It closed in 1969. FR STU 10/105. Courtesy of Museum of English Rural Life, University of Reading.

By the turn of the century opportunities for women to formally train in agricultural education had begun to emerge. Once the case for agriculture as a "profession" for women had been won, the hierarchical system of agricultural education remained, embodied in three tiers: universities providing three- or four-year degrees, agricultural colleges offering two-year diplomas, and farm institutes giving a range of short courses. Although most of these institutions were open to women, the number enrolling was small. More important were the specialized women's colleges that provided agricultural and horticultural training. Swanley Horticultural College admitted women beginning in 1891 and became a women-only college in 1902.[49] Studley College was opened in 1903 by Lady

Women students in a peach market at Swanley Horticultural College, Kent, UK, before World War I. Swanley opened in 1887 and began to admit women students from 1891. In 1902 it became a women-only college. Between 1891 and 1914 it charged a fee of £70 per annum for training and was aimed at the daughters of middle-class metropolitan families. SR 2WFGA PH5/46. Courtesy of Museum of English Rural Life, University of Reading.

Warwick and was the successor to her earlier venture, Lady Warwick College at the University of Reading, which ran from 1898 to 1902. The women's colleges specifically attracted the daughters of wealthy metropolitan men and in a direct social parallel with the campaigns of Langham Place, recruited those who wished "to enter the profession, not those born to it." Their fees effectively barred those from below the middle classes.[50] These colleges were also assisted by associations formed specifically to help women train for farming careers. The Agricultural Association for Women, founded in 1899, published the instructive but short-lived periodical *Woman's Agricultural Times*, which ran an employment registry and marketing department among its other activities. In the same year, the Women's Agricultural and Horticultural International Union was formed (later becoming the Women's Farm and Garden Association). Each was an inheritor of the Langham Place call to arms and a sign, as Sutherland concludes about other avenues of middle-class women's work, that change was really starting to happen, not that it had been fully accomplished.[51]

The Langham Place group recognized the value of skilled, knowledgeable farm food production and the rational, reasoned choices of rural working-class women to put both that knowledge and their labor at their families' disposal. The group's understanding of farm food production as a profession fit for middle-class women who had had no previous experience of it, yet who might learn it not just abstractly but also through practical engagement with its processes, took agriculture itself beyond the realm of "vernacular" knowledge to address agriculture as a science and a business that belonged to the world of skill and economic production. Thus, they rewrote British middle-class femininity to include embodied, experiential knowledge. These ideas remapped women's participation within the countryside, challenging dominant constructions of class for both laboring and elite women, and established the need for agricultural education. The Langham Place feminists should be seen as the intellectual forerunners of a later group of women whose interventions were more practical, but who were themselves still bounded by conventions of respectability and femininity. The Langham Place feminists are significant because they opened up the debate about the place of women in farming in the 1860s and 1870s, using forms of respectability that included but also leaped over the association of food production with domesticity. In their promotion of agriculture as a suitable profession for middle-class women, they uniquely set the agenda that began to reap results a generation later in the era of the New Woman.

2

Good Farms, Markets, and Communities

Emily Hoag and Rural Women as Producers

JOAN M. JENSEN

Early in November 1920, thirty-year-old Emily Hoag, an assistant economist in the U.S. Department of Agriculture, left Washington, DC, by train for North Carolina to launch a research survey of rural women. Eventually, this survey would take her beyond North Carolina to South Carolina, Virginia, and West Virginia, then north to New York and Massachusetts. In all, she would be in the field more than eight months, traveling from farmhouse to farmhouse, some days on foot, other days by mule team or horseback and, thankfully, often by train and automobile. In one letter, the director at the Office of Farm Management (OFM) at the USDA which sponsored this research, reassured a farm woman, "we know that Miss Hoag does not mind walking if necessary to reach places on her list (we know of one instance already where it was necessary for her to walk seven miles)." In another letter to a prospective interviewee, the Agricultural Extension Director for West Virginia described Hoag as "a plain sensible country woman whom you will be glad to know."[1]

In one sense, Hoag was a "country woman." Raised on a farm in upstate New York, the stout pair of walking shoes that were a requirement for this research would have been familiar to her. But in other ways she was a New Woman, part of that late nineteenth- and early twentieth-century cohort of women who sought more independence, visibility, and participation in public life. Emily Hoag was also seeking a place in the newly forming agricultural professions in which men were calling the shots or, as some scholars have put it, forming "the new patriarchy." This essay focuses on Hoag as a new woman in the emerging fields of rural economics and sociology, part of a group of reformers who hoped to redefine women's role in country life as well as their own role in the study of it. She was not a "home economist" but an agricultural economist who studied women's work in the farm enterprise. This analysis of how Hoag negotiated a career within the emerging field of rural studies is based on a combination of family archives, public records, Hoag's lively field letters written while employed by the USDA to study rural women still preserved in the National Archives, and her final unpublished report "The Advantages of Farm Life." Her place in the newly evolving agricultural professions was intertwined with what historians have called the "country life movement."[2]

The country life movement was an effort to restore country life and the production of food as an important factor in the modernization of the economy in the United States. Many participants in this movement also thought the old ideal of the white yeoman farmer as the moral bedrock of American identity was being threatened because so many native-born white youth were abandoning the countryside for the seemingly more exciting life in the city. Yet the burgeoning cities still needed a large country population to provide their food. To keep young men down on the farms, it was necessary to keep young women there. Redefining the production of food as a partnership between rural women and men offered one way to make rural life more attractive to young women. Farm women would no longer be "helpmeets" to yeomen farmers but partners in the modern production of food and the policies that guided it. Hoag's experiences are an example of how women who desired public professional agricultural careers in the United States navigated within these shifting gender attitudes during the early twentieth century. Her research demonstrated the importance of women in the production of food and food policy.

Hoag prepared for and made her country journeys of 1920 at the dawn of the fields of rural sociology and agricultural economics, part of what we today call rural studies. Then, seemingly, everything had to be created. Not only did these new agricultural professionals develop goals and methods, they also shaped and defined the social hierarchy within which they would work. Hoag was one of the first women to participate in creating the disciplines of rural sociology and economics. In those early days, there was some question of whether this gender and racial stratification would be reworked for the new fields of knowledge.

When Hoag arrived in Washington in 1920, the USDA classified her as an assistant economist, making her the highest ranking woman in the department. Whether Hoag was to be a harbinger of women's new role in policy making or an aberration would be determined during her next six years at the USDA. For the preceding three years, Hoag had worked for the agricultural economics department at the University of Wisconsin. During her years at the university, she had provided important empirical data to support theoretical work, including what came to be known as the "agricultural ladder." Hoag's painstaking research with the farms near the university seemed to prove that tenant farming, then a growing concern among agricultural economists, could provide a "ladder" to farm ownership, still the accepted ideal for farmers in the North and Midwest. Even though this optimistic view of tenancy conveniently omitted dealing with African American farmers in the South, it provided much-needed reassurance to Euro-Americans that the older pattern of widespread landownership by white farmers was still viable despite growing tenancy during this period of rising prices for land and farm equipment. Hoag's research in Wisconsin helped convince agricultural economists that the older farm ownership practice could survive the modernization of United States agriculture.

To survive, however, the whole concept of what it meant to be "rural" in modern life had to be reevaluated. As the urban population swelled during the early twentieth century, defenders of this emerging urban culture increasingly denigrated rural men as "bumpkins" or "selfish brutes" and rural women as victimized drudges of these men. Such attitudes embellished the urban myth of progress and the backwardness of the life people from the countryside left behind for "modernization." In popular literature, rural people had become just another group of "exotic locals" to be satirized, joked about, and dismissed. These people not only had "problems," they were "a problem." In response to this negative view, a romanticized view of the more distant rural past developed. It had been good, now it was bad. Rural sociology and economics emerged as specialized social sciences devoted to explaining changes occurring in the countryside as the United States industrialized and modernized. Later, male academic scholars seldom mentioned women such as Hoag who had worked beside them in those heady days when their fields of research were taking form. In fact, women were central to the modernization of agriculture and to the scholars who studied it. This chapter considers how one exceptionally intelligent, well-educated, and experienced country woman navigated the waves and shoals created as rural sociology and economics took form, and how she found a place within them.[3]

The Making of a "Plain, Sensible, Country Woman"

Born on a farm in rural upstate New York in 1890, Hoag first traveled east to Albany for her education, then west to work at the University of Wisconsin in Madison, and then southeast to Washington, DC. During the first thirty years of her life, she emerged as a New Woman able to get college training in New York and then moved up the educational hierarchy, finally in North Carolina teaching others how to teach rural students. In 1917, she accepted a job as a research assistant at the University of Wisconsin in Madison (UWM), then one of only three universities offering courses in rural sociology and agricultural economics.[4]

The United States had just joined the Allied war effort when Hoag arrived in Madison. At UWM, as on other college campuses, both male students and faculty were soon absorbed into war-related work. When rural sociologist Charles Galpin accepted wartime work, Hoag took over his courses, and when the war ended in November 1918, she stayed on as his research assistant. She carried out field research in the surrounding farm country and assisted Galpin in transforming that research into articles and books. She would continue to work with Galpin as assistant economist in Washington, DC, at the USDA's newly formed Division of Farm Population and Rural Life, where she researched, wrote, and, as it turned out, completed the only study of farm women to be sponsored by the government during the first half of the twentieth century.

When Hoag arrived at the USDA in 1920, she became part of a group of well-educated women already hired as specialists. While the upper echelons of the Department of Agriculture bureaucracy were still staunchly male, women had increased their presence during the 1910s. As assistant economist in the OFM, Hoag was the highest ranked woman, but others were in executive and administrative positions, worked as research specialists, and served as secretaries to bureau chiefs, auditors, and librarians. Other women worked with Hoag in the USDA: eighteen in the Bureau of Chemistry, seven in the Bureau of Entomology, five in the Bureau of Animal Industry, and fourteen, including two women with doctorates, in the Bureau of Plant Industry. There were zoologists, pathologists, mycologists, and even a xylotomist who conducted research on sweeteners. By 1920, the secretary of agriculture boasted: "There seems to be practically no limit to the positions which women may fill in this department." There were enough women working in the Washington office, he wryly added, to get a no-smoking rule adopted for the building.[5]

Educating a "New Woman"

Still, Hoag's long journey from rural New York to Washington, DC, and her rise to assistant economist in the USDA, is worth examining in greater detail because it was part of the attempt by political progressives to broaden the opportunities for professional women and to assign those with rural backgrounds to carry out research related to the production of food.

Emily and her two sisters grew up on a rented farm in Granville, Washington County, New York. She later wrote fondly about life on that Granville farm: "We grew up in a lovely country, with beauty on every side, and were always making wreaths and garlands for ourselves, dandelion curls, burdock baskets, telling fortunes with daisies and sticking butter cups under people's chins to see if they liked butter." Although poor, she wrote, her parents did not seem to know they were poor, and so the three daughters did not realize it until they were older. The family took in wayfarers. The tin peddler and the clock tinker who made them toys would bed down in the attic and eat with the family. There were always books in the family and pride in the distant relative Horatio Alger, who wrote optimistically about how poor boys could rise to middle-class security through hard work and the help of wealthy patrons. By 1910, despite these hard times, the family had moved from tenancy to ownership, perhaps because the daughters' off-farm incomes provided the family with enough economic security to become landowners and to hire a young man as a live-in laborer.

The Hoag Broome County farm was outside the town of Windsor, just north of what is now Highway 17, the main road from Binghamton to the Catskill Mountains. That part of Broome County had made a painful transition from wheat farming to dairying in the nineteenth century. Then, because it lay along the Susquehanna River with access to water transport and railroad connections, Windsor partially industrialized. The Hoags lived on East Windsor Road, north of the village and near the river. Windsor had a railroad depot, post office, church, and school, but not much else. Farm women there faced grim conditions—scattered farms connected by unpaved rutty roads. If a family had no kin living by, as was the case for newly settled families such as the Hoags, it was hard to socialize. The women who helped operate family dairies had little time to socialize anyway. They usually had no hired help to assist in their expanded dairy work, and they continued to assist in growing diversified crops and subsistence gardens, all the while caring for small children and the farmhouse. Except in summer when some families took in boarders, they had little female company. Boarders meant more work for the farm woman, but at least offered company for a few months out of the year. Life was so hard that many families simply abandoned the area. After touring the area in 1909, the secretary of agriculture described the upland areas as an "abandoned farm area." Two years later, after a similar tour, President Theodore Roosevelt's Country Life Commission reported that there were few takers for the many farms for rent at low cost.[6]

Moving to that farm on East Windsor Road was probably a good move for the Hoags. It meant that the daughters were able to find teaching jobs in the small villages, which had difficulty attracting teachers. With the incomes earned by teaching daughters, the family was able to buy land for the first time. Teaching was one of the few ways for farm daughters to work off the farm, and they were welcome as boarders everywhere in these isolated farmhouses. For daughters raised on these struggling farms, the path to teaching lay through classes at a normal school. With financial help from her sister, Emily was able to earn a BA and an MA at the New York State Normal School in Albany and to prepare to teach in a high school. Hoag's first job was teaching English and history at the high school in Ovid, a small village near the Finger Lakes in New York, but she soon moved on to teach at the normal school in Asheville, North Carolina. She taught psychology, sociology, and education, did social work among the Appalachian mountain people, and ran a model school for students preparing to teach in rural communities. When not teaching, she continued to take classes herself—at Peabody College in Nashville, Tennessee—and attended Chautauqua lectures, which were a popular way for rural women to continue their education. She absorbed the progressive agenda for the advancement of rural people through embracing scientific farming and education. She wrote a pageant, *The Mountain Maid*, and with the help of the local YMCA and community actors, presented it in May 1916.

About 1914, Emily Hoag posed for this formal portrait in Asheville, North Carolina, where she was teaching at the normal school. Three years later, still in her early twenties and restless and eager for new challenges, she resigned her position in Asheville and moved to Madison, Wisconsin, where she began her research on rural life. Courtesy of Donald W. Sawtelle Jr.

It began with Sir Walter Raleigh planting seeds from America in Europe and ended with teachers preaching the benefits of scientific farming.[7]

Now a young and restless twenty-six-year-old with years of experience teaching, writing, and administering, Hoag continued to look for a way to hone her own skills and connect her own rural past to the current issues of rural life. She consulted with a teacher in Albany, who urged her to apply for work at the University of Wisconsin in Madison. In 1916, the University of Wisconsin was assuming distinction as J. R. Commons, Richard Ely, and H. C. Taylor worked out the theoretical and methodological underpinnings for the new social sciences, especially sociology and economics. The sociology of rural life was just taking form under the energetic leadership of Charles Galpin in the new Department of Agriculture. Although Madison was the state capital, when Hoag arrived, it was still a small college town located in Dane County, surrounded by farms fanning out on all sides.[8]

As the state land-grant university, UWM's mandate was to teach agricultural courses. In 1904, President Charles Van Hise had reformulated the school's mission as the "Wisconsin Idea," dedicated to meeting the needs of the state's citizens. With the establishment of a graduate school, older students began to arrive to study, not just from Wisconsin, but from as far away as New York and Maine. The College of Agriculture had an ambitious outreach program, offering short classes to farm sons and even a few daughters, and an active extension program which pioneered in exploring and attempting to help solve what had come to be known as "rural problems."

The agriculture college at UWM had been swept up in the country life movement. Two concerns preoccupied the agriculture faculty—the public perception that the rural population was being left behind by the urban population and that, infested by problems, farm people might turn to radical political and economic solutions to solve them—or even worse, just abandon farming and cease to provide the food necessary for an urban population. Northern Wisconsin was settled by predominantly European immigrants who, as lumbering gave way to farm settlements, moved from mixed farming to production of butter and then cheese. These farms were already industrializing. Dairy farms were everywhere, acting as milksheds for the newly urbanizing population and selling their cheese and butter on the national market. Down in Madison, the agriculture college specialists were worrying about the reluctance of farmers to abide by new health standards as they shipped milk to nearby cities, as well as their seeming inability to maintain a consistently high quality of the butter and cheese that they sold cross-country. Rural reform, to deal with "rural problems," followed the commercialization of farm production. As a land-grant university, UWM had a mandate to serve the state's rural population, but it also had a mandate to serve the economic needs of the state. For both goals, agricultural experts needed more systematic studies of rural life, both the people and the economy. UWM had one of the first departments of agricultural economics. When Hoag applied for a job in the Agricultural Economics Department, it had just hired Charles Galpin to teach a new class that applied sociology to rural areas. Like Hoag, Galpin was from upstate New York, had been educated in country schools, and had taught in a country town. After moving to Madison in 1901, Galpin became a sort of lay minister to rural students, helping them organize a club. Through the club, he met H. C. Taylor, head of the Agricultural

Economics Department, and was drawn into the discussions of rural living conditions and what came to be called the country life movement.[9]

The country life movement reached the University of Wisconsin campus most visibly when, in 1909, President Theodore Roosevelt's Country Life Commission arrived for hearings. The faculty in the newly formed Department of Agricultural Economics listened attentively to the commission's lament that rural life was becoming a "problem." Young people were fleeing, depopulating the countryside, while those left on the land were not competing properly in world markets. The experts reasoned that if rural life could be made more attractive to these youth and if they could be trained to take advantage of new industrial opportunities for farm production, they might stay on the farms. The agricultural economics faculty, including Galpin and department head Taylor, listened attentively, and as Galpin later wrote, this "gospel of a new country life" had converted him to studying the "social forces" at work in rural lives.[10]

During these years, Galpin and the faculty of the agricultural economics department in Madison developed the academic disciplines of rural sociology and agricultural economics. Some academicians, like Galpin, were committed to "social Christianity," or what historians call the social gospel movement. Instead of immutable natural laws, these reformers proposed an "intelligent direction of social change," what economists now call "institutional economics." Reformulated as secular and scientific, the movement retained its goal as the welfare and well-being of the population but proposed training specialists who could offer scientifically based reasons for change. The practitioners of this movement planned to test theories about agriculture and rural life by gathering information from and about actual farm communities. Looking back at the development of this new "social science," Galpin would later write that reformers first approached rural conditions like missionaries intent on improvement and conversion to new agricultural practices, then identified specific problems, to which, once identified, researchers could formulate solutions to systematically eliminate these problems. Scholars interested in reform now had an agenda. They just needed to identify specific rural problems.[11]

When Hoag arrived in Madison, Galpin showed her his method of identifying problems by "mapping" rural social institutions. Using data from Belleville, the little New York village of six hundred where he had taught for sixteen years, and enlisting the aid of the Belleville librarian, who spent three months collecting data, Galpin created his first "map."

The map showed that tenants who lived on back roads had far fewer social contacts than owners who lived near main roads, convincing him that geography had much to do with social conditions. Organizing this first study and mapping the data, Galpin later wrote, "did a thing to me which came well-nigh to making me burn my books and stop going to libraries." It turned Galpin from a convert into a preacher with a method for his own message. In January 1911, the University of Wisconsin hosted its first statewide conference on farm problems. With Taylor's support, Galpin presented a series of rural maps at the conference, published them, and then used them to teach his first class on rural life. It was one of the first courses in what came to be known as rural sociology, the systematic academic study of rural communities. Like any new field, Galpin had no textbook, so he used published reports by the Country Life Commission and the Madison conferences, as well as his own study, as models to train students to do field work.[12]

Hoag assumed, as did Galpin, that a systematic knowledge of the human and social side of farm life had to be a part of understanding rural life. She also agreed that the human situation in farm life would yield to a scientific method of inquiry, based mainly on the survey. As yet, however, there was no consensus among researchers in this newly developing social science about how to design such a survey, gather data, and interpret the results. In preparation for a speech at the December 1920 meeting of the American Sociological Society in Washington, DC, Galpin circulated drafts to a number of scholars working in rural studies. He believed that every fundamental human situation in farm life would yield to scientific methods of inquiry and that the main method was the survey. The question was how to formulate and carry out the survey. The replies to Galpin's draft revealed two important questions: how to select the topic of the survey and to what extent the people to be interviewed should be involved in that selection.

Experts themselves disagreed. E. C. Lindeman in the Department of Sociology and Economics at North Carolina College for Women questioned whether there could even be a "purely scientific" survey unless specialists cooperated with the community from which they gathered their data. Howard W. Odum, director of the School of Public Welfare at the University of North Carolina agreed that it had to be a "combination of scientific and community spirit." Margaret F. Byington, who in 1910 had authored the groundbreaking sociology study *Homestead: The Households of a Mill Town* and was now director of field services for the American Red Cross, also thought community participation was essential. And Liberty Bailey, now seventy-two and retired from heading the Cornell College of Agriculture but still an important leader in the country life movement, insisted to Galpin that the investigator should interpret as well as collect the data. Another important question was the role that the community should play in determining what was to be studied. These were major issues for the social sciences that did not exist in the physical sciences, tensions that would also be built into Hoag's major research survey on rural women.[13]

When Hoag reached the Department of Agricultural Economics at Madison in 1917, she would have immediately been immersed in these questions about the goals and methods of the emerging field of rural sociology. Galpin had adapted the scheduling method of mapping urban institutions to the study of rural social and economic institutions. Scheduling involved listing questions that could be answered quickly through brief interviews, which could then be combined with census material and other documentary materials to assemble maps. Galpin considered three thousand filled schedules adequate to yield significant information and used this method for his first study, "Rural Social Centers in Wisconsin." Galpin's assumption was that a small study of a rural community, Delavan, could be used as a proxy for the larger Walworth County and that this method could then be used to describe even larger areas of rural life. Galpin used this research as a basis for short courses and semester-long classes on rural life, small graduate seminars, and advanced research projects for graduate students seeking higher degrees in rural sociology and economics. What he termed "rural sociology" was, in fact, rural life in Wisconsin. Meanwhile, he wrote articles on rural life for the *Country Gentleman*.[14]

When Hoag met Galpin in the spring of 1917 and became his research assistant, he was the most important scholar interpreting rural life. Galpin's prolific writing combined with his fervor in preaching the need for systematic study of rural life gained him a reputation as a leader of the new rural sociologists. At first, Hoag helped him write up his research, but after the United States entered the European war that April, most of the male students and many of the faculty at UWM were drawn into the war effort. Too old to serve in the military, the loss of his male students simply gave Galpin more time to write. Gradually, ways for men his age to help the war effort at home developed, so he joined the state guard. Hoag, twenty-seven years old and an experienced teacher, capably took over his classes, collected data on rural conditions, and organized research results for publication.

It is important to remember that Hoag was entering a university hierarchy that had only recently integrated its female students into full equality in education. For years, women had battled for that equal place at the University of Wisconsin. Between 1875 and 1900, women had gradually been able to integrate the university, moving from a separate female college to access to regular classes by 1900, when women were half of the graduating class. They had been accepted, as Amy Hague wrote, but "encouraged to accept a narrow definition of what they could do with their lives." Conditions for female students changed after 1900 as the university began to grant advanced degrees to women. Few of these women would assume prominent public positions. Most used their expertise to support men in rural sociology and agricultural economics. But a few were ready to assume positions of responsibility in their own right. Like Hoag, many of these "new women" professionals were former teachers and administrators with years of experience. While women students had a chance to see a woman teaching during the war, when the war ended Hoag returned to being Galpin's research assistant.[15]

By September 1918, as the war drew to a close, Hoag had made her first survey with a "farmstead to farmstead" visit of five hundred farms in Dane County. Farm tenancy was one of the major issues being debated in 1917, and at the American Sociological Society meeting that year members had proposed that field studies of tenancy be conducted in every state. Hoag's research in Wisconsin was to become the prototype and best known of these tenancy research studies. The data Hoag gathered became the basis for a well-known article on the "agricultural ladder," authored by Galpin and sociologist Richard T. Ely and published in the *American Economic Review* in March 1919.

In a footnote, Galpin and Ely merely mentioned Hoag as the assistant who gathered the data for the agricultural ladder article. A separate UWM bulletin, authored by Galpin and Hoag, described her research methods in gathering data for that article. First, she located with a dot each farm in four "townships," the designation used in Wisconsin for counties. Then she chose Sun Prairie with its 1,200 inhabitants and gave each farm a number (1–500) on her survey sheets. To make her visits "easy," Galpin announced in a local paper that Hoag would be visiting farms to ask about past occupancy for the years 1909–1918.

Hiring a horse and buggy, Hoag systematically visited each farm home and asked: "Who has occupied the farm in the last ten years?" If the occupants did not know, she quizzed the neighbors. The answers showed that a large number of farmers had moved from tenancy to landownership. In their article Ely and Galpin used the statistical data Hoag had gathered in support of their theory of the "agricultural ladder." Tenancy was not a bad thing; because tenants moved to landownership in large numbers in some areas, tenancy could be a positive step in acquiring land. They offered this small piece of Wisconsin research as a standard of measurement for an ideal system of landownership. Hoag was now an essential part of Galpin's ambitious plans for the field of rural sociology and the place of the "Wisconsin school" in it.[16]

The Wisconsin school had now come to dominate emerging rural studies. Not just their theories, but the scientific research that claimed to support the theories, had thrust Taylor, Ely, and Galpin to the top of an academic hierarchy, catching the attention of U.S. Secretary of Agriculture David F. Houston, who shared the ideals of the country life movement. Houston offered Taylor a chance to set up the Office of Farm Management (OFM) in Washington, DC, which would apply, on a national scale, studies of farm economics similar to those done for Wisconsin. Within the OFM, Galpin would have charge of a farm population and rural life division and a committee on farm life studies to suggest priorities. This twenty-eight-member Committee on Farm Life Studies, which included five women, recommended that the new OFM expand the small amount of specialized market research already being done to the farm as a whole and its relationship to communities and the nation. It identified the farm home as the fundamental unit of rural social organization and the committee's first priority, with an initial focus on the "farmer's wife" and children within that home. It wanted studies of women's attitudes, work, their relation to farm income, and their interest in management, health, and contacts off the farm. This was an astoundingly new direction for agricultural policy. Galpin asked his indispensable research assistant to accept a position as assistant economist in this rural life agency. Hoag joined Galpin in Washington, DC, in the spring of 1920.[17]

Learning the Ropes in DC

It was an exciting time for a young professional woman to be in Washington, DC. During the war, women had taken over many of the government's bureaucratic duties. As men settled back into peacetime occupations, many women were able to move up within the bureaucracy and abandon clerical work, just as they had abandoned teaching in the late nineteenth century. Coming from rural New York and Wisconsin, where few African Americans lived, the biracial composition of the capital might have been new to Hoag. She had, however, apparently lived comfortably in segregated North Carolina. She would have easily fit into the white population of the city. The District of Columbia was not as rigidly segregated as other Southern cities, but residential segregation was strictly enforced. Hoag's boarding house lodged white women. Black women served them. Such segregation lasted well into the 1960s, when desegregation attempted to end this practice. Hoag quickly found a place to stay at one of these newly formed women's boarding houses, where she met other women like herself, eager to join the federal postwar government agencies. By the time Hoag arrived in Washington, Galpin was already busy planning tasks for his rural studies research team. He inherited a female clerical staff of three—a secretary, a head clerk, and a typist—who would have helped Hoag find her way around this new "women's Washington."[18]

In an address before the American Home Economics Association, Galpin tried to explain just how the new OFM would fit in with the already well-established field of home economics. Home economics, he explained, studied the physical basis of the rural home, while the new farm life studies would take the "life side of the farm home" as its focus, primarily the "state of mind of the members of that home" and the problems arising in the home. It would analyze the social relationships on the farm, not its physical basis. He sketched his conception of farm home life as an "economic partnership" of the adults and children. The goal of the partnership, however, was to raise the son to manhood and establish him as the replacement of the father and thus insure the survival of the family cycle, the welfare of its apprentice child and soon-to-be manager, tenant, co-owner, and complete owner who, in turn would continue the cycle and pass it on to his son. As for the woman's role in this cycle, rural studies would study the "spiritual" role of the woman in the home as the interpreter of one family (her own) to another (her husband's) and of her children to their father and the father to his children. Agricultural work was a "common task," involving the cooperation of the whole family. Thus, all phases of farm life needed to be studied. Hoag's job was to find out how the distribution of tasks led to a "contented farm home" and the willingness of "superior" young women to remain in farming.[19]

While Galpin did not describe Hoag's research, we know that in June 1920 she was just embarking on a major study of rural women that would last four years.[20] Galpin was clearly using the techniques they had already worked out in Wisconsin: find a data set, plot it on a map or graph, and write it up in the form of a bulletin. Hoag finished her four-year study in 1924. That report, "The Advantages of Farm Life: A Study by Correspondence and Interviews with Eight Thousand Farm Women" was never published by the government. It proved to be the only study of rural women by the OFM. In that study, Hoag concluded that farm women belonged to three important sectors: the farm as a joint business enterprise principally producing food, the home as a second economy in itself, and the community through their role as activists involved in the welfare of others. Thus, Hoag did not see rural women as only having a role in what was coming to be called "home economics"; rather, they had a much broader role in the economy and the culture.[21]

Finding a Data Set

Both Hoag and Galpin knew they needed a statistical sample that would reflect the larger cohort of farm women they wanted to study. The first task, then, was to locate a data set that could throw light on these issues and help them determine policy. Small studies told them little about exactly what role gender played in rural life. They had pieces from Wisconsin, New York, and Illinois, and scattered studies from the South, but nothing that isolated women specifically. There really was no data from which to fashion such a study—or so they thought. While Hoag finished up her New York migration study, Galpin kept looking for a data set suitable for research on farm women.

Hoag and Galpin eventually found that several sources did exist. One source was some two thousand replies from women to a 1913 USDA survey. That year the new Secretary of Agriculture, David Houston, sent out fifty-five thousand letters of inquiry to farm families across the country. He asked no specific questions but encouraged women to discuss any need that occurred to them. In the 1913 USDA yearbook, Houston summarized the results in "The American Farm Woman as She Sees Herself." Houston divided the responses into four topics: social and labor needs, domestic needs, education needs, and economic needs. These responses had offered officials a "bird's eye view" on the life of farm women, he noted. His was the first attempt to consider women as a part of the USDA's mission to benefit the agricultural population.[22]

Houston noted in this summary that the responses indicated that programs already in place had not succeeded in reaching farm women, especially in the South. What they wanted was available, but not to them. Their hunger for assistance of all types seemed to amaze him, for he noted that these requests ranged from help in the care of babies to information about operating gasoline engines. They had, he wrote, "unbounded confidence" in the ability of agricultural and domestic science to revolutionize their lives. But women wanted that information not through the printed page, even if available, but in lectures, movies, exhibits, and demonstrations, something the recently passed eight-million-dollar Smith-Lever Act would seek to supply. With "a touch of bitterness," he noted, women wanted homes to be considered a part of the management of the farm business. With the declining availability of farm labor, women had less assistance in the home and were also being asked to share more of the work outside the home. They often suggested the expansion of cooperatives, not just for buying, selling, and processing, such as in creameries and cheese factories, but also in labor cooperatives—laundries and bakeries—that could be developed alongside processing plants.[23]

The women did not seem to want to spend more time in the house but less, to have time to develop livestock and poultry enterprises, something the USDA also endorsed. As the size of farms increased, the lack of workers and cooperatives made women more isolated on farms now located farther apart along poor roads. Women were concerned about the poor schools as well. They wanted better schools, but not at a distance, and schools that helped educate children to stay on the farm rather than prepare them for urban life. Mitchell, and probably the women, worried that as farmwork became more demanding for women, their daughters were coming to despise housework as "degrading drudgery." Continuing education, or what was called extension service, seemed to be an important solution. From his analysis of these letters, Edward B. Mitchell concluded that rural women wanted not only better schools but also more cooperatives to help with their tasks. They wanted better libraries, meeting places, and leadership through cooperatives and women's clubs. These women writers did not consider themselves as "a class apart" but as part of the general community. They wanted healthier and better lives for themselves and their families and believed that the government could help them through economic legislation and better education.[24]

These two thousand letters, along with the Mitchell report itself, became a part of Hoag's research data. In this source, the voices of rural women in the South were especially muted. Not only were the voices of the black Southern farm women conspicuously absent, but even those of white farm women. This silence led Galpin and Hoag to make a special effort to interview white Southern women.[25]

A second data set soon became available to supplement the Houston letters. In September 1919 the Eastern edition of the journal *Farm and Home*, then published in Springfield, Massachusetts, printed two articles, "The Woman Favored by God" and "The Woman God Forgot." As was often their custom, they asked women readers to write to the editors about their responses to the articles, discussing how they viewed women's rural life. Galpin had frequently written for *Farm and Home*, and in November he wrote to the head of the Western office in Chicago, Herbert Myrick, to ask if he could borrow the responses for a study. Myrick referred Galpin to the Eastern office headed by Edwin C. Powell, who then forwarded several hundred letters, which formed a second data set for Hoag.[26]

The Interviews

Hoag's approach was to base her research on these letters and questionnaires, supplemented, crucially, by her interviews with farm women. Galpin had already decided on some sort of work in the Southern states before he obtained the letters from Powell. Because the magazine had primarily New England and East Coast readers, interviews in the South would not only give Hoag direct contact with rural women but also expand the geographical scope of the study. Galpin and Hoag decided that she would conduct one hundred interviews to supplement the letters, beginning immediately with a short exploratory trip to North Carolina, South Carolina, and Virginia in November and December. This trip would be followed by a longer field trip in 1921.[27]

By the fall of 1920, Galpin was corresponding with the presidents of colleges with agricultural programs and extension programs to ask for their cooperation in selecting one hundred "exceptional farm women" with whom Hoag could visit. In her report, Hoag described the data set as being composed of these interviews, together with the Houston letters, the Phelps publishing letters, and a new batch of letters from rural women collected by the Webb Company, based in St. Paul, Minnesota, and publisher of the *Farmer's Wife*, the largest periodical aimed specifically at rural women, with a readership primarily among farm women in the North Central states.[28]

Hoag was well into her interviews when the Webb letters came to light in March of 1922. The letters offered a broad view from a different region. In December of 1921, the editor of *Farmer's Wife* published "Do You Want Your Daughter to Marry a Farmer?" It invited farm women to reply to the question by writing about their own lives. Of the more than one million subscribers, seven thousand women, mainly in the Midwest, took the time to respond, thus enabling a tabulation of information for that region.

As he had done with the earlier Powell letters from *Farm and Home*, Galpin now asked the Webb Company for the use of these letters for Hoag's study. Bess M. Rowe, field editor for *Farmer's Wife*, sent the Webb letters to Galpin in the spring of 1922. In a letter to Galpin, Rowe wrote: "Probably never before has the real heart of the American farm woman been so fully expressed as it is expressed in these letters." The evidence was overwhelmingly in favor of farm life, she concluded, and the letters that were not in favor came from areas of the country where agricultural conditions were especially hard or from women with "individual tragedies in their lives." Hoag now had some eight thousand letters from farm women in addition to her growing notes from interviews.[29]

Most of the women who self-selected with their letters and the rural women chosen by extension agents were white and, although some were not well off, most seemed to consider themselves to be middle-class. Not all of them were happy and satisfied, as comments in Hoag's letters to Galpin indicate. Some ranked themselves very low on the happiness scale. Letters and interviews also revealed a number of different ways in which those who were achieved that feeling with their life styles.

The interview phase of the study of rural women, as it eventually evolved, included home visits to white farm women in Southern and Northern states. In the winter of 1920 and spring of 1921, Hoag visited four Southern states—North Carolina, South Carolina, Virginia, and West Virginia; between January and April 1921, she also made shorter trips to Massachusetts and New York.[30]

Galpin explained Hoag's project in correspondence with agricultural experts stationed at land-grant colleges, telling one that this was to be "a great study of the optimism bearing upon the situation of women in the problem of country life." Something must be done soon, he went on, "to overcome the tremendous amount of pessimism which the press is putting out." Although the easiest way to reach rural women was through home demonstration agents already in the field, he cautioned Hoag to only use them to contact women and not include them in her interviews. "I shall get along without their help as much as possible," Hoag promised.[31]

Hoag reviewed the criteria for selection of women in a letter to Galpin. The Greensboro, North Carolina, demonstration agent had selected "progressive" women for interviews. All were involved in home demonstration activities, but their work arrangements were vastly different. One woman and her husband were raising pure-bred livestock together; a second farm woman did most of the farmwork because her husband worked off the farm. On a third farm, the husband shared some of the work, but as a school principal he was also absent part of the year. On a fourth farm, four unmarried sisters shared the tasks: two brought in cash from teaching jobs, a third managed the house, and the fourth handled all the outside work. Thus, some thrived on field work while others seldom went out of the house or garden to participate in either production or marketing of farm products, and some worked at whatever tasks were at hand. In her final report, Hoag smoothed over most of these differences in order to present one happy group.[32]

Neither Hoag nor Galpin made any effort to include poorer white farm women in the sample. Nor did they include rural black women, some of whom had carved out successful farms in the aftermath of the Civil War and had continued to farm, as the black abolitionist Frances Harper explained after touring the South in the 1870s to 1890s, with "energy and executive ability." After 1904, Jeanes teachers—sponsored by Anna T. Jeanes, a Quaker benefactor—left scores of descriptions of successful farming projects started by black students and their families. Under the 1914 Smith-Lever Act, black extension home demonstration agents and their clients worked with women farmers. Hearing of Hoag's interviews with farm women, Anna Garrett, superintendent of Negro schools in Columbia, South Carolina, asked whether the USDA wanted to interview colored women in Richland County. Apparently, she received no reply. There is no indication that the USDA even considered including black women in their survey. Hoag's letters mention only one conversation with a black Southern woman in North Carolina, when she casually asked a black woman in domestic service why she was not working on a farm. The answer was brief: work too hard; pay too little. The story of black rural women farmers is told more fully in Cherisse Jones-Branch's essay in this volume.[33]

Given this narrow slice of white farm women, what do Hoag's unpublished letters to Galpin reveal? In one letter from West Virginia Hoag praises the farm women as "wonders." But, Hoag adds, they were by no means all happy or optimistic. Among the interviews of five women, two were optimistic, two pessimistic, and one "downright tragically unhappy." Overall, she had found four happy women, each of whom Hoag found to have "a great vision and a great motive," with the enthusiasm and perseverance to work out her desires. "So long as she has the strength and opportunity to work to fulfill those desires, she asks nothing better," Hoag concludes.[34] As she started the New England interviews, Hoag found the farm women in New England less effusive, more restrained and "bashful" than those she interviewed in the South. She wrote from Massachusetts: "I am finding it a little difficult to penetrate the New England reserve." She also found the state "filled with rich people's estates" and spent only two weeks in the state before returning to the South for another six weeks of interviews.[35]

By the time Hoag began her visit to upstate New York farm women, she had already conducted more than four-fifths of the planned one hundred interviews. Galpin began the search for New York farm women in the spring of 1921 with a letter to agricultural economist Dwight Sanderson at Cornell University's College of Agriculture. Sanderson drew up a list and asked the women to participate, then Galpin passed on the names to Hoag, who wrote to confirm the time for the interview. For New York, perhaps with so many Cornell contacts available, Galpin selected twenty instead of the usual fifteen women for Hoag to interview. Hoag's two months of travel mainly by train would save her shoe leather and energy. Hoag was on home territory now.

Hoag's letters to Galpin from the New York farms she visited were full of suggestions for reforms based on what the women had told her. Instead of planning programs, agents should ask women what they wanted. It was true that women had much to do, Hoag wrote, but women had a tendency to hide behind work. "I know, because I have been there myself, and lived among them," she wrote, "it's masquerading, all the same." By reorganizing work schedules and adding some labor-saving devices, the farm women would have more time for other nonfarm tasks, home schooling their children, self-improvement through reading, and even writing about their farm experiences. Hoag suggested a circular recommending writings such as those by New York humorist Marietta Holley and encouraging rural women to tell their real stories for publication.

Individual self-improvement for farm women was only one of Hoag's suggestions for the use of their time. They should use some of this newly found time to improve their communities through developing ways to work together. She noted with pleasure the building of a community house at East Handsfield and one at Adams. They might be able to do other things: establishing relations with urban customers might help expand the demand for farm-grown products and help the customers understand the costs of producing food.

In her letters, Hoag described many of the women she visited that summer. Three New York farm women impressed her especially: Cola L. Fountain, northeast of Waterton in Evans Mills, Jefferson County; Mabel G. Feint, northeast of Ithaca in Dryden, Tompkins County; and Mrs. Alfred Abbuhl of Greene, northeast of Binghamton, in Chenango County. Each of these women had written a reply to Sanderson's request for an interview, and he passed the letters on to Galpin and Hoag. These three letters survive in the archives, along with Hoag's letters describing her visits to them.

Hoag found in Cola Fountain the rural life philosophy she had been seeking. A local Watertown paper had first published Fountain's stories, the State Home Bureau had reprinted them, and Hoag had read them. Fountain sent a carefully typed letter from Evans Mills expressing interest in getting "hold of the constructive side of the question" of farm life and welcoming a visit from Hoag. She explained that she could not be considered either as a successful or representative farm woman, for they had "much to be desired financially," but she would not trade farm living for a million dollars. She closed her letter with the note: "I haven't been outside of Jefferson County but once in 13 years, and I am always glad to meet someone who can widen my vision to something beyond the churn and the butter-bowl."[36]

Hoag described Mabel G. Feint as another ideal farm woman. In a long handwritten reply to Sanderson, Feint had welcomed a visit from Hoag but warned that she did not live in a "high type of the comfortable farm home." She went on to explain that she was not strong but could no longer afford to pay wages for a hired man and had to do all her own housework. Two years earlier, she had suffered a physical breakdown from overwork, then opted to continue community work, both locally and beyond, and Faint confessed she no longer attempted to keep a systematic or careful house but only provided for the basic comfort of the family. It takes, she went on, "every ounce of energy I have to keep going." Her home was "badly in need of repair" and had few conveniences to make a guest comfortable: no pipeless furnace, no water system, no toilet conveniences. They had the money for the bathroom equipment, but with no hired help, her husband was so busy outdoors that he had no time to put it in. Instead of installing the bath system, she had asked that he devote any spare time to work in the kitchen. Her chief "business" was to write on commission for farm and dairy papers, which she was doing that week, in addition to canning. If, knowing all this, Hoag still wished to visit, she was welcome to come. "I do believe most heartily in the possibilities of a very satisfying, wholesome sound life in the country," Feint concluded, "but I also believe that few in higher official circles really understand the hardships and handicaps that prevail on the average farm." Should Hoag attend a local Home Bureau meeting and survey the "better class" of women who attended, Feint suggested, she would find that few had home conveniences and that the women of "humbler families" could not even afford to join the Home Bureau. Every woman was invited, but "after one or two visits pride will keep most women from going to a club she cannot help to support financially." Among club members, only a few had conveniences; most did not. Feint closed with instructions on how to reach her home—four miles outside of Dryden off the state road from Courtland to Dryden, on Gee Hill Road, with the road in fairly passable condition.[37] Hoag visited Feint early in July, presumably over passable roads. Feint was perhaps not the "ideal" woman, but one who was willing to tell Hoag what even women most committed to rural life were facing. They acknowledged the facts of their farm lives and a willingness to wait for household conveniences.

If Fountain was the perfect person to work for improvements, and Feint the cold voice of reality about the inability of rural women to obtain them, then Mrs. Alfred Abbuhl (she never used her birth name in identifying herself) proved to be the farm woman with the most perfectly balanced philosophy of life. In January 1921, Abbuhl had penned a long reply to Sanderson's request for an interview with Hoag. "I am *not* a successful farm woman," she began her letter, writing that she was often not optimistic, had no views or experience that would be helpful, and that a visit from Hoag would be a waste of time. But, Abbuhl continued, she would be delighted to have a visit from Hoag. The idea of meeting Hoag was thrilling, and the experience of meeting her would be an "inspiring and interesting and encouraging experience." Abbuhl offered a moving description of her attitude about farm life: "I love the farm and the farm problems and the problems of the farmer's wife. I know the life in the country teaches us to appreciate the best, the real things and puts us in touch with that which is beautiful and wonderful and soothing. I would hate to leave the farm." Her wish for farmers, she concluded, was to have "*convenient*, comfortable homes, their own autos, the cost of production plus a reasonable profit for their products & a little time for recreation and money enough so they can come to Cornell for Farmers' Week *every* year."[38]

Sanderson passed on Abbuhl's letter to Galpin, who sent her a copy of his recently published book *Rural Life*. In a letter to Galpin, Abbuhl recounted her life as a rural woman. She was a grandmother, wanted to contribute to the social life of her neighbors, and also to get one of the farmhouses he suggested in his book. She had already organized a farm and home club that sponsored a community picnic among other activities. Rural people, she concluded, were starved for a larger life. In another letter, she recounted her day: washing, attending to cows and chickens, working in the garden, baking, doing housework—fourteen hours of hard work. Abbuhl still argued that the visit was a mistake, for she was not successful or intellectual, and at times not optimistic. Still, her farm was paid for and she lived there with her three daughters, all of whom she had put through high school, and one through Cornell College. But she had no modernization on the farm, and she longed for "life." By that she did not mean material things, but spiritual things. She had enough material things in the birds, flowers, weeds, brooks, and the silence of rural life. It was good to wear old clothes. If, knowing all this, Hoag still wanted to visit, she was welcome to come.[39]

When Hoag visited Abbuhl about a week later, Galpin's book was on her reading table, a rose bush bloomed at the front door, and a collie was at her side. Abbuhl bragged about her three grown daughters. After the visit, Hoag wrote to Galpin that Abbuhl was a remarkable woman whom he should visit someday. Hoag planned a return visit with a home demonstration agent, promising she would surely find an optimist and, perhaps, some ice cream. When the two arrived, Abbuhl served them banana ice cream and sponge cake.[40]

Hoag's "The Advantages of Farm Life"

With the completion of her New York interviews in the summer of 1921, Hoag's field work was done. During the next year she formed her interviews and the thousands of letters from farm women into a compact research report. Her goal was clear. Farm people resented their "characterization and cartooning as ignorant objects of misguided pity." She would show farm women as they were: "strong, resourceful, capable, and leading personalities in their communities."[41]

Hoag's study has been important to rural historians for documenting the work and attitudes of rural women in the early 1920s, but the question of how rural women fit into the formulation of food policy by the USDA and the United States as a whole has been underestimated. Hoag identified not only the important place women occupied in the rural economy but the importance of including that role in determining agricultural policy. Such a bold claim could not go unchallenged for two reasons. First, it moved women into a place where their work in producing food could not be ignored. Second, it assumed that this knowledge of women and their role in producing food should be acknowledged in formulating agricultural policies. Such conclusions cast her study as a part of a major debate over agricultural food policy as a whole. Hoag and the other Wisconsin economists viewed the economy as a reflection of economic policies that could, and should, be formed by institutions that took into consideration the effect of the economy on the rural people involved in agriculture. This evolutionary view came to be called "institutional economics." Neoclassical economists, however, argued that the market was not to be tampered with except to help business flourish. As we know, those two economic theories are still being debated almost a century later.

With progressives in charge of the USDA during the early 1920s, Hoag's positioning of women in the rural economy and in the production of food seemed to reflect an important reality. But if that reality were applied to food policy, then women should also be involved in formulating that policy. And including women in the formulating of policy, at a time when they were on the cusp of obtaining the vote, was certainly a challenge to the old days when men could create policy without considering all the producers of food. In fact, the 1920s were to become a battlefield on which the secretaries of agriculture and commerce contested the boundaries of food policy. Thus, the decision of how to shape studies of rural life and women's place in them was a political issue almost from its beginning.

Hoag arrived in Washington just in time to see these various debates work themselves out politically and conceptually. One debate was whether to keep economics and rural life together in the USDA or move economics to the Commerce Department and abandon rural life studies. Another issue was the place of economic studies of women. One view was to put such studies entirely within the emerging field of home economics and the home extension divisions that were formed through the implementation of the Hatch Act of 1887, which had established extension programs at land-grant colleges. As colleges and departments of home economics took form within agricultural colleges, the field became the choice of many women as well as men who wanted to emphasize women's roles as consumers and homemakers and not to have to deal with male supervisors. An alternative was to keep the study of women within the departments of agricultural economics as a part of rural life. If women were to remain within rural life studies, then the third debate became relevant, whether to keep the study of rural women a separate study within rural life or to merge studies into the category of "farm families."[42]

The question of women's role in rural life had been on the agenda of the rural life movement almost from the beginning. Women had been at the core of its problems and its promise. As half the rural population—two-thirds if children were added to their mothers—what women did and thought was important to social and economic issues. As half of the newly eligible adult voting population of 1920, when women's suffrage became law, women were potentially an important voting bloc among the burgeoning urban and declining rural populations. The fate of rural women—or their "condition" as it was often termed—seemed in public debates to be a factor of major importance in policy decisions at the beginning of the 1920s.

The men who fashioned the early policy of the Office of Farm Management believed that rural life was a combination of both social and economic interests and activities. Julie N. Zimmerman and Olaf F. Larson, in their study of how the agency took form, called it a "holistic" view, one that took as its purview a broad view of rural life that included agricultural production of food but saw that production as a part of a larger culture. The farm, the home, and the community were all part of this life. While the incorporation of modern farming methods into rural life was a concern of both the agriculturalists and the rural life proponents, there was yet another issue. If women's labor on the farm were to be improved, perhaps even lightened, how would she use her new freedom? Should she devote this time to leisure, recreation, and self-development? Should she become a better homemaker, consumer, and mother? Should she seek new ways to economically benefit the farm enterprise? Or should she devote her time to public policy, at the local level, through community affairs? Another view was possible, of course, that not only should and did she work in the house but she also worked in the barn and in the community, organizing to support both house and barn. And she decided when and how much time she devoted to herself.

Hoag's statistical tables from the farm women's letters remain in the archives. They were accompanied by a twenty-nine-page USDA typescript summary of her survey, dated March 1924. She did not include a description of the process by which she integrated these thousands of letters with her more than a hundred interviews to form a coherent report. Hoag did acknowledge her sources: the letters from the Webb Company in St. Paul, Minnesota, the Phelps Company in Springfield, replies to USDA questionnaires, various publications of the Country Life Commission, and her own visits and letters to farm women.

Hoag's appendix located the letters as coming from rural women in the North Central region of the country. Most were of moderate means who had lived more than ten years on farms (92 percent), with a large minority (43 percent) who had also lived at some time in the city. A majority were farm daughters who performed farmwork (two-thirds), and half of the women considered themselves "working partners." A large percentage of the letter writers, more than 90 percent, were generally satisfied with their lives despite discouraging times. Many did complain about the lack of conveniences in the home, overwork, and poor schools, but these concerns were not limited to just farm women. At least a quarter of all the writers complained that farm people in general were referred to as "rubes" and "hayseeds," and rural women misrepresented as drudges immersed in "hopeless and meaningless toil." Instead, they described themselves as proud workers worthy of respect and admiration.

Hoag divided the report itself, now titled "The Advantages of Farm

Life: A Study by Correspondence and Interviews with Eight Thousand Farm Women," into three equal parts: work, social, home. She began with work, which she described as based on love of the land, citing Booker T. Washington and an unnamed Wisconsin Native American scholar. Then she cited the importance of farmers' weeks at agricultural colleges, which emphasized not just home economics but also offered lectures on all that went on at farms and reforms that were important to both farm women and men—better roads, consolidated schools, cooperative creameries and laundries. Hoag described farms as partnerships to which a woman contributed through "her garden, her butter, her hens." The farm enterprise was an interchange between house and garden, barn and field. She cited Southern practices in which women managed the "little crops," the peanuts, potatoes, corn, and nuts, the diversified food crops that kept a balance on the farm, while the men managed the "big crops," the single market crops, often cotton or tobacco. She emphasized the work of children and the desire of farm families to have higher education for their daughters so they could support themselves. She agreed that there was a need to do away with drudgery but that the caricatures in the press had it wrong—there were vexations and hardships, not drudgery. She quoted one woman who spoke of her "sisters on the farm" and their love of the soil.

This partnership and love of the land kept women on the farm. Their enterprises settled communities, which then became an integral part of national life. Hoag sketched a female view of the frontier, dismissing as "pre-frontier" the migration of men, detached and adventurous, who worked alone, and created a way of life that suppressed and replaced the social side of life as the center of attention. This attitude had made rural life "unsatisfactory, one-sided, incomplete, unstable, almost a soil-mining venture for the purpose of obtaining money with which to return some day to more congenial atmosphere and mode of life." She went on to argue that rural women's "creed," while emphasizing that making money should not deprive her family of social privileges, and that she be "a courageous social pioneer as well as a resourceful agricultural frontierswoman." She described history in terms of stages: pioneering, primitive civilization, modern organization. For Hoag, modern organization was logical and good, for rural women wanted access to concentrated, clusters of services. Farm life, she wrote, was now a trial to demonstrate whether farm life could pass through this "era of organization." The "thinking farm woman" believed that nothing could prevent modern organization of farm life but did not believe in blind submission to it. Hoag quoted pointedly from a letter that claimed that women wanted public policy and institutions that would allow them to have a happy fulfilled life in occupations they loved. Good land, good markets, good communities. Women wanted to build community life along with building economically successful farms.

After work and community, Hoag analyzed the "home side," where, as one woman wrote, the farm home spread all around. This was not a structure or the work there but a larger metaphor that encompassed the entire nation. Hoag was anxious to note, however, that this was not based on pure tradition or custom but on a lively farmhouse that sheltered and encouraged reading of all types, from adventure books for children to current events, not just "chickens and fertilizer" but the opening out of larger vistas. She lauded the rural rather than the urban home as defining the nation. This part was easy, quoting from others who had praised the farm as being home to the whole nation, the "rock-bed of patriotism," the "stabilizer . . . to make us a strong and permanent nation."

This was a large vision of farm women, stretching as it did from partnership and farming to community, and including the farm home, "the rock-bed," the "stabilizer." Hoag had distilled out the discouragement and depression that she encountered and wrote about in her letters to Galpin and created a huge role for rural women in the economy, social policy, and home life. This was not a balanced or scientific view. It seemed to have little relationship to those tightly defined county studies that were being developed by other rural sociologists. But it had much to do with the Wisconsin and New York progressive vision of the potential for rural life. What she had done was to put women at the center of that rural vision. In doing so, she linked rural women to the aspirations of their city sisters as well as to each other and to the men on the farm. She made a place for rural women in the future, not just as a part of history.[43]

Farm Women Become Home Economists

Despite the efforts of suffragists to mobilize women behind progressive reform policies, including those that made rural women an important part of agricultural policy, Democrats lost that 1920 election. It is a complicated election to analyze, but one result was that while the OFM continued to exist within the USDA for some years, it shifted attention away from the overall conditions of farm women to their roles as farm wives and mothers, thus folding women back into the family and their work as part of home economics. Following the defeat of the Democrats in 1920, Galpin stayed on at the USDA, refocusing the direction of the OFM but never quite solving the split between working bottom up from the needs of the rural people in formulating national farm policy and the top down, fitting rural people into the needs of the national economy. Taylor left the OFM and returned to the more quiet waters of academic politics in Madison. Secretary of Agriculture Edwin T. Meredith founded a women's magazine that came to be known as *Better Homes and Gardens*.

In 1920, Meredith bragged, justifiably, that he had placed women in new positions of responsibility in the USDA. For a few years, progressive Democrats had brought new women, such as Hoag, into higher positions in the USDA. Nevertheless, although Hoag completed her massive study of rural women during 1922 and the study was finally prepared in March 1924 as a bulletin, it remained unpublished. The progress Meredith had celebrated a few years earlier was not sustained.[44]

To explain that lack of momentum for Hoag's work and the work of women like her, it is worth noting that the USDA was undergoing a change of direction. One last activity of Hoag's for the USDA occurred in 1922. The USDA *Official Record* listed two lectures given by Hoag, the first in March at Bryn Mawr College in Pennsylvania on "rural life" and a second in June at a teacher's summer institute in Castine, Maine, on "rural sociology." The titles, perhaps significantly, indicate a shift from rural studies that included agricultural economics to rural sociology, a move that located their research firmly among the emerging social sciences and within the male-dominated academic institutions.[45]

At the same time, the USDA signaled that it would focus more narrowly on home economics. Carolyn M. Goldstein, in her analysis of home economists, explains a shift in USDA policy. In 1923, the USDA established the Bureau of Home Economics (BHE), evidence of its shift from interest in women's role as producers to their importance as consumers. Now, with their own bureau within the USDA, home economists expanded their role as experts who, as Goldstein explains, mediated the relationship of women to the marketplace. In other words, Hoag's view of rural women as producers and community activists lost out to the view of women as primarily consumers and homemakers. The task of women agricultural professionals was to explain and promote these roles. Home economists had not only organized nationally to protect their new professional status but also created a place within the universities and the USDA for them to practice their expertise.

Writing Women Out of History

Those lively letters that Hoag wrote to her boss, Charles Galpin, in the 1920s were what first interested me in Hoag's work and the role of rural women studies in the USDA. Although Galpin had recruited Hoag to join him in Washington as his assistant economist and organized her study of farm women, when he later wrote his history of rural sociology, he did not describe Hoag's work or the contribution of any women, either as researchers of this new field or as subjects of research. In Galpin's history, women and gender did not play a role in the creation of rural sociology and economics. There is no mention of the "new women" who assumed public roles as reformers or of the role they played in what historians came to call the "country life movement." Yet the country life movement derived much of its power and popularity from the participation of women who were active in every stage of its existence. As interest in women's role in rural life increased, women were gradually added to country life committees and to research projects. Women were included in the first OFM committee, which met in 1919 to establish priorities for research. They were also an important part of the new field of sociology and economics, although male academics soon separated sociology from social work and economics from home economics. What had been an early interest in "women" as a separate category of rural research was soon subsumed into "rural families" and home economics.

The rise of home economics as an academic discipline and as part of the USDA explains, to a great extent, why Hoag disappeared from history. In 1923, just as Hoag completed her study, the USDA established its Bureau of Home Economics. The elevation of home economics to a separate bureau within the USDA reflected a shift in interest from rural women as producers to all women as consumers of farm products. Home economists were to be the experts trained to study these women. Home economics found a place in agricultural colleges, and women who wished to study rural life could choose to study the "family" in rural sociology or home economics in agriculture. The "place" that most women students occupied in land-grant colleges and their "schools" of agriculture became "home economics."[46]

Recent studies by Grey Osterud, Mary Neth, Katherine Jellison, and Julie Zimmerman have all described the importance of farm women to rural economies and communities as well as within the farm home. This research has shown that Hoag's broader view of women as not only important in the home but also in the business of farming and in community activities did not disappear in practice. But no national movement equal to the home economics movement emerged to perpetuate and expand Hoag's broader role for rural women. After 1923, Hoag's views gradually disappeared from the USDA.

So then, if Emily Hoag's brief public life does not represent the way USDA work was trending, why study her? Because she was the first woman to have a place in the emerging fields of rural sociology and agricultural economics. She had half the rural population to deal with, almost no relevant methodology or tools with which to study them, and little encouragement to continue her work. Her study of rural women, although unpublished, is the only systematic study of rural women conducted in the early twentieth century and remains an exception to the invisibility of women in rural studies. Her reports to Galpin, those exuberant letters from the field, testify to the excitement of this rural daughter, this New Woman, over the possibilities of studying women as a separate and important half of the rural population, partners in the farm business, the basis of home life, and critical participants in rural community life. That is enough.

3

Professionalizing Farm Women, Recognizing Their Integrated Food Roles

The Netherlands, 1880–1950

MARGREET VAN DER BURG

A farm household management teacher once said, "If it is about development, you do not talk about nutrition only, you also speak about health, food and agriculture. Many other issues must also be attended, such as clean water, hygiene etc."[1] Her words characterize the targeting of farm and rural women following agricultural and rural development initiatives in the Netherlands in the 1860s. These initiatives were developed in response to international and national debates in rather progressive agricultural, women's, and religious circles. Because family farming was predominant in Western Europe, the question of how women would be taking part in the modernization of agricultural production and rural life had to be addressed. Searching for policies suitable for the Dutch rural areas, advocates of agricultural reform experimented with various ways to professionalize women, particularly family farm women from various types of farming, farm sizes, and regional cultural and religious backgrounds. Educated women would be the "new farm and rural women" who would act as role models to other women in their families and communities. Reformers expected especially to find women leaders for these initiatives among the daughters of the better-off farmers and rural professionals who had advanced education and a feel for farm and rural circumstances. These women, once educated in training institutions for teachers, could take up leadership and delicately maneuver in the midst of male agricultural and rural opinion leaders, religious authorities, and urban women leaders.

The reformers viewed improving food security and safety as an integral part of agricultural innovation and rural development. The overall goal was to secure an economically viable agricultural sector, together with a healthy rural population and modern living conditions that would keep them from leaving the countryside. Farm women's contribution to this goal included both securing a healthy farm family and producing quality food safe for commercial purposes. Besides gardening, processing meat and vegetables, and preparing meals for household use, many Dutch farm women were involved in the farm labor of milking, watering cattle, dairy processing, feeding the chickens and pigs, weeding, haying, and harvesting fruits, potatoes, and wheat. According to many spokespersons, to perform these tasks optimally, women required education and learning so that they could select and work with new technical options provided by the agricultural sciences.

By highlighting the context of agricultural and rural development initiatives in the recognition of farm women's integrated food roles and how they changed over the years, this chapter adds a new dimension to food studies. It approaches food studies from a rural gender perspective that goes beyond solely urban household consumption and also beyond male-oriented agricultural production by putting the focus squarely on the family farm household. It thus expands the diversity of contexts by giving attention to farm family households where food is produced, preserved, and prepared. It therefore enables a new integrative, gendered analysis of these aspects of food and respective women's food roles.[2]

Targeting Women in Agricultural Innovation and Rural Development

Around the middle of the nineteenth century, European elites became interested in agricultural innovation and rural development. They started to work together with their governments through new societies that saw their task as either to encourage welfare in all economic sectors (welfare societies) or to concentrate specifically on agriculture and rural development (agricultural societies). Around the 1900s, farmers also started to promote their own interests by forming farmer's organizations, syndicates, or unions. Like similar societies in North America, these organizations promoted innovation based on new scientific insights gained through experimentation and systematic data collection.

At first, the agricultural societies were composed mainly of the rural gentry: better-off farmers, rural entrepreneurs with agricultural estates, businessmen involved in agricultural industries and trade companies, and rural professionals such as notaries, headmasters, priests, and ministers. Some single and widowed women of the same class also participated; some became rather well known for their reports about their international journeys and visits to conferences all over Europe and beyond.[3]

In addition to growing interest in the unhealthy conditions of urban working classes, social liberals and democrats also became interested in the living conditions of rural people. Although "peasants" had previously often been portrayed in a romanticized way as living the "good, natural" life in harmony within their natural environment, social realists started to counter this perception. For instance, Vincent van Gogh's *The Potato Eaters* is a wonderful example of paintings portraying the hard work and living conditions of rural people.

Regardless of these sympathetic portrayals, the "enlightened" upper and middle classes with bourgeois aspirations continued to depict peasants as "others." They newly framed them as rough and backward, kept in the dark, deprived of new findings and insights. Consequently, such people urgently needed to be elevated or "civilized" to better lives. Their ignorant existence should be enlightened by education that combined scientifically based knowledge about agriculture with general education in the spirit of patriotic and religious virtues.

At the same time, educated, better-off farm women and rural gentry "ladies" started to extend their domain in philanthropic work. A.M.M. (Mienette) Storm-van der Chijs (1814–1895) is a noteworthy example. She was one of the first Dutch feminists fighting for women's equal access to professional education and careers, specifically in agriculture; thus she was also a New Woman herself. In her speeches and writings from the 1860s onward, she distinguished two groups of women in agriculture who could profit from the upcoming social change. Better-off young women would be enabled to provide their own livelihoods by taking up professions in agriculture, such as dairy processing, horticulture, poultry keeping, or beekeeping. She demanded agricultural education for these women that was practice oriented but also scientifically based. Farm wives and mothers, she argued, needed an education that would support them in better mastering their work and responsibilities. In her opinion, these women needed advanced elementary education combined with new insights from the sciences applied to their specific responsibilities as farm women, wives, mothers, and housekeepers. This would help them elevate themselves from being not just "first laborers" on the farm but also leaders in bringing the rural population to a higher moral standard.[4]

Construction of the New Farm Woman within Family Farming

Both options presented by Storm-van der Chijs were soon being discussed at international conferences. Later, information on new initiatives was exchanged at the international conferences of the International Council of Women (started 1888) and the International Congress on Agriculture (started 1889).[5] While the first feminist initiatives were solely oriented to women's domains in agriculture, severe rural circumstances and growing rural out-migration led agricultural and rural opinion leaders to promote education that covered all the work and responsibilities of farm women.

This last approach reflected that farming was mainly family farming, as it existed and still exists in most of Europe and certainly in the Netherlands. In family farming, however, men were considered the professionals par excellence; women were generally denied this status. Demanding education for farm women to professionalize them indeed called for acknowledging their roles on the farm. But according to the dominant family ideology of the time, there was a place for only one "farmer." Agricultural debaters quickly underlined the complementarity of women's work but did not address issues of equality, such as a redistribution or sharing of tasks or overall responsibilities. Nevertheless, these new educational ideas had to acknowledge how much farm and family were intrinsically interwoven. Women's work for the farm, family, and the house often overlapped and conflicted because women had to attend to many tasks at the same time and switch from one to the other. It also meant that any change on the farm would likely affect her work too. Nevertheless, the debaters never fundamentally questioned the gender-specific rules and arrangements that largely constituted family farming. Examples include the succession and management of the family farm, labor arrangements, and the distribution and control of resources.

Paul De Vuyst (1863–1950), inspector of agricultural education in Belgium, had a strong impact on policy makers and was key to the consolidation of the new approach toward the education of women in agriculture in many countries. His writings of 1891 and 1907 reflect in great detail information he collected on existing initiatives concerning farm and rural women.[6] First published in Flemish and French in 1907 and translated into English in 1913, his most important book, *Women's Place in the Rural Economy*, highlighted the kind of education that should be further developed to strengthen farm women in their social role in rural development.[7] Meanwhile, De Vuyst arranged to have a resolution accepted at the 1907 International Conference on Agriculture in Vienna that obliged all countries present to discuss the establishment of agricultural education for farm women.[8] His ideas were preceded by earlier initiatives that called for further investigation on how these should be connected, especially with regard to the U.S. extension service and Canada's Women's Institutes. He made particular reference to these efforts as examples of what had been done to educate rural women in other countries.[9] His speeches and writings inspired Netherlands agricultural and feminist leaders to critically discuss which policies regarding women's education might suit Dutch circumstances.[10]

Ultimately, the construction of a "new farm woman" emerged from these discussions. It embodied a new concept of rural and farm femininity within which women's specific roles on the farm could be combined. The new farm woman would still perform farm labor, as well as provide food for the family through the production, preservation, and storage of foodstuffs. She would combine work in the house and on the farm, challenge existing farm and rural life conditions, and participate in rural social and cultural life. She would bring up the children and prepare daughters for their futures, preferably in service of modern family farming.

Establishing Special Women's Education within Agricultural Education

There had been earlier attempts in the Netherlands to establish special education for women in agriculture. In the 1890s, when women still dominated dairy work, special classes were set up to train them using new methods and equipment. Soon after, however, dairy factories began to take over the dairy tasks. Only in the region where the famous Gouda cheese was made did the women continue making the cheese on the farm, and so their education continued.[11] In addition, a private horticultural school for girls, Huis te Lande, was started in Rijswijk in 1907.[12]

Around this time, women's rights advocates, agricultural organizations, and governmental officials joined into the debate to discuss how foreign experiments might be adapted to rural conditions in the Netherlands. National agricultural conferences in 1905, 1911, and 1924 all addressed the topic at the national level, and leaders urged agricultural organizations to discuss what to do at local and regional levels. In 1905, the principally male participants agreed on the need for extended education for farm women on economic grounds. They argued that improvement of the core activities (in other words, farm men's domains) on the farm could only be successful if the secondary trades and the housekeeping (farm women's domains) got full attention as well. This meant that farm women had to be educated in the wide field of their activities just as men were educated for their work. Provincial organizations reflected the same attitudes in their meetings. In Gelderland, one leader proclaimed that the need to provide such education was self-evident, a view that was sketched out colorfully in 1909 by a spokesperson in Limburg:

> Who looks after and feeds the cows, no, all the animals, big and small, down to the chicken? The woman. Who consequently gets involved in the purchase of concentrated animal feeds? The woman. Who nurtures the calves and piglets? The woman. Who milks and takes care of the dairy? The woman. Who generates a lot of distress when the home garden does not produce? The woman. Who has to care for the food of the family? The woman. So, would it not be very useful that our farm daughters get educated on the care and feeding of cattle? Livestock? Dairy? Fertilization and efficient cultivation etc. of the garden? Simple book-keeping? The preparation of a simple, cheap and still nutritious meal?[13]

By this time, a general agreement in the Netherlands had been reached on establishing a new type of vocational education for women in agriculture that put their work and life circumstances in the center. Besides the direct importance of their agricultural labor, women's indirect economic impact on the farm accounts was acknowledged; first, through running the household and, second, as the "educators par excellence" by maintaining high moral standards for husbands and for the next generation. Neither the existing agricultural education oriented toward men's agricultural labor nor household schools oriented toward urban domestic life were suitable.

Veendam opened the first state course in 1909. Literally translated, the Dutch name *landbouwhuishoudonderwijs* means "farm household management education," and because it also included agricultural education, it cannot be translated simply as rural home economics education. It started out as a women-oriented part of agricultural education and developed in the 1920s into a regular branch of education for girls and young women in the agricultural sector. After World War II, it became the educational training for the majority of rural girls.

Securing Diversity in Farm Household Management Education

Dutch social democratic agricultural opinion leaders and policy makers were determined to reach out to as many farm and rural women as possible. Religious agricultural organizations, however, challenged what farm women should learn. In any case, all agricultural and other organizations were entitled to apply for state support for the women's courses, just as they were for the men's. This meant that policy makers and educators were urged to deal with many diverging opinions, initiatives, battles, and coups. Thus, there was great diversity in educational practices, but there were also ongoing discussions and negotiations about what education would be best. State inspection enforced some uniformity by maintaining specific principles in their criteria for funding. State programs had to report back, which enabled national administrators to keep track at the national level. This way, the educational programs could be monitored and expanded coherently. In the end, the various programs reflected their negotiated views on women's part in agricultural modernization and rural development.[14]

Different levels of education were clearly defined and reconsidered as necessary. Some agricultural organizations offered extended part-time courses in schools in regional small towns. Others created elementary part-time courses to rotate around villages. Often organizers combined both approaches. Wherever possible, administrators made arrangements to use specific equipment and classrooms in existing agricultural or household schools. They catered to villages with rotating courses at local primary schools, community centers, and even in pubs. The southern Catholic farmers' organizations embraced the concept of boarding schools and completed them with training farms on the school grounds. The Catholic organization in the eastern part of the Netherlands estab-

Farm daughters in chemistry class, farm household management course, Kollum, province of Friesland, 1920s. Learning not only how but also why new methods would be used through chemistry and physics lessons was considered crucial for all courses. The course was held in the household school of a small town; other times, a room of an agricultural school could be used. Courses at a school had a more extended curriculum than rotating courses in the villages. The photograph shows rather young girls from middle-class farm families who could afford to miss a daughter in the farm and farm household work for some days a week. Rijksarchief Overijssel, inv. 435.1, Zwolle, Netherlands.

lished a traveling school to serve the families of smallholders. A common policy was enforced, however; each curriculum had to reflect farm women's tasks in the region and class hours had to be adjusted to the time women could be away from home duties. Classes were scheduled during the least busy periods of the year. Regionally, there was a clear variation in age groups, but schooling was mostly reserved for young women aged fourteen to sixteen years or older.

Leadership by New Women Teachers for Agricultural and Rural Development

To cover all the curriculum subjects, teachers from various professions were asked for their support. Primary school teachers with an agricultural teaching qualification taught the agricultural subjects. Regional family doctors or district nurses took care of health instruction. To take on leadership and address the rural life conditions or rural home economics, special farm household management teachers were to be trained. At first, home economics teachers with a farm and rural background were attracted, but later assessments concluded that these teachers were not well enough equipped for their tasks. To fill this gap, the state established an experimental training school in 1913. Six years later, in 1919, the school was made permanent. A few denominational training schools were opened in later years. Theda Mansholt (1879–1956), a home economics teacher and farm daughter, was appointed principal of the state-training school. She succeeded in getting newly trained women teachers appointed on a par with the male state agricultural agents or consultants. Farm or rural background was set as the decisive criteria for their selection.

Because these teachers had to be leaders as well, they had extensive tasks in addition to teaching. They were supposed to provide extension services through farm visits, give lectures, stimulate the opening of new courses, and also to advance farm women's own continuous learning. They promoted the use of general agricultural media while also creating specific media for farm and other rural women. These included magazines, brochures, notebooks, leaflets, radio programs, and market and exhibition stands. They also urged former pupils to organize themselves into associations that would help them continue learning, organizing activities, and advancing their interests.[15] The teachers offered their support for those activities by setting up libraries, helping promote the establishment of medical care services, the installation of central water pipe lines, and the construction of an electricity and telephone network.[16] Thus, the female teachers became key professionals in monitoring farm and other rural women's development work.

This leadership position required a broader curriculum than was needed to only give instruction in rural-specific home economics. Student teachers learned not only about all farm women's labor domains but also the socioeconomic and cultural differences between all regions in the Netherlands. In addition to learning theory, they had to do practice, both in the house and on the extensive home gardens. Students rotated specific chores every month. They learned to keep household budgets similar to those of farm families. One of the former students underlined the significance and appreciated the attitude behind it: "Poor and less-educated people were never spoken of in the usual protective way. On the contrary, we learned to have respect and admiration for these courageous women who had to do their work under such difficult circumstances."[17] This observation corresponds well with Theda Mansholt's idea that whatever the type of education, it had to contribute to the people's well-being: "the household has to be seen as an *organic whole,* a business that is run in the interest of the family members, thus of *people,* and that cleaning, cooking, the laundry, make no sense for the cleaning, cooking and washing in themselves, if they have not been done in the genuine interest of people."[18]

Threatened and Well Secured in Addressing Women's Farm and Rural Identity

This farm women's education was formally brought under the Directorate of Agricultural Education. Both its strength and vulnerability were soon tested under the attempt to transfer it into the new Act for Vocational Education (Wet op het Nijverheidsonderwijs) in 1919. This new act would not include agricultural education but fuse farm household management education with other women's domestic science and vocational education. It would also be transferred to the Ministry of Education, away from men's agricultural education. Advocates of farm household management education feared that the curriculum would lose its specific agricultural and rural orientation, and thus its importance, if the education was to be combined with urban domestic science education. One of the farm household management teachers voiced this concern: "In the life of a woman on a farm her tasks as housewife and woman farmer are so much interconnected that one cannot—or can only with great difficulty—separate them, except perhaps at larger farm estates. Therefore agricultural education is an integrative part of farm household management education, and this education will never meet its goal if both tasks are not well addressed."[19] One teacher-to-be argued that failure to maintain a separate identity could mean the end of having farm women coming to classes because: "The protest by the farm household management teachers might be successfully overcome, but never the silent protest of the farm population. Education that is not serving their needs, will simply not be attended, it will bleed dead."[20]

Echoing this sentiment, Parliament member Van Lanschot moved a resolution through Parliament in 1919 to give formal recognition of farm women's specific status, arguing: "All of this education is permeated by the idea that these pupils have to become farm women; it makes its specific character. Not only agricultural lessons are given in these schools; also household education is oriented toward the practice of farm women. One does not have to be an educational expert to understand that household education for girls who will live on farms is quite different from household education for girls who will live, for instance, in a city, in a middle-class family."[21] The result was that a specific identity for farm women's education could be upheld. Instead of being merged with urban household education, it remained a separate branch of vocational education for girls. The continued involvement of agricultural organizations was also ensured. In the 1930s, farm household management teacher Greta Smit (1900–1980) was appointed as woman inspector for this branch specifically.[22]

Food and Nutrition: Beyond Cooking Like Their Mothers

Food and nutrition were an integral part of the women's curriculum. Farm women played a decisive role not only in producing food for commercial and family use but also in the processing, storage, and preparation of food. Farm women's accounts illustrate how much food-related work defined their daily lives, especially on smallholdings in regions with sandy soils, where women reported that "they were the first to rise and last to go to sleep."[23] At the agricultural conference of 1924 a member of parliament pointed at the importance of acknowledging this reality in farm women's education by ridiculing the urban household school curriculum, saying "an average Brabant farmer needs a companion who knows her way in the garden and with the chicken more than one who has recipes for ten different puddings in her head."[24]

Nevertheless, women's reluctance to come to classes had to be overcome. Women were used to learning women's work from older women. Teachers often heard the refrain, "Cooking I can learn from my mum."[25] But when neighbors or family took the first step, others wanted to be part as well: "If sisters and cousins had followed a course, or friends, then you went there too. You wanted to belong."[26] From the beginning, teachers used a variety of techniques to attract the attention of their potential pupils to farm household management classes. For instance, before the start of a new course, the farm household management teacher might give a lecture at a meeting of the local agricultural organization. At one of these lectures in the province of Friesland, more than two hundred people showed up and far more pupils than could be served inquired about new classes. Agricultural exhibitions and fairs also proved to be excellent places to attract attention and interest.[27] In the province of Drenthe in 1938, one teacher reported: "All we taught was demonstrated. . . . Success we had! A lot of interest from the Drenthe women, who expressed their admiration for all those wonderful fruits in glass jars: 'How is it possible, how is it possible' one could hear all around. Effect: full classes the coming years!"[28] Special events also stimulated pupils to widen their scope and perform in public. In 1920 the Gueldern organization held a special Farm Women's Day on its seventy-fifth jubilee.[29] In Purmerend in 1928, the local agricultural auction cooperative asked for help with introducing a new product—the tomato: "In our department we had a long table on which we cooked and placed the prepared dishes at the end for display, and another table where visitors could sit down to taste the various dishes. There were a lot of inquiries, and the pupils showed again and again how they had cooked what was tasted."[30] Other teachers arranged for their pupils to cater at agricultural events and festivities.[31] They also organized excursions for their pupils to visit an exhibition, event, factory, or institute related to the course.[32]

Especially for the rotating part-time courses, the teachers had to find out what would work best for their food and nutrition lessons. At first, the cooking was mainly taught through demonstrations. Tasting helped overcome fear of something new. Learning by doing it oneself was preferred, however. Special transportable cooking equipment containers to be folded out to cook on were developed for use in villages that had no well-equipped school. The equipment could be set up and put away again, and then transported to another village when the course was finished.[33]

Seasonal vegetable canning was often part of a program. Special arrangements with a local butcher were made to encourage safer and more nutritious meat processing. Schools in small towns could offer other options as well. Although it took them a long time to convince municipal administrators sometimes, teachers were eager to introduce

Farm daughters practicing poultry and gardening lessons, farm household management school, Lonneker, 1920s. Gradually, more specifically equipped schools were established. Most had two or three rooms with at least facilities to practice cooking and laundry, but a school vegetable garden and even poultry to practice what had been learned were on the wish list of many women directors. The school at Lonneker served as one of the advanced examples. The one at Zierikzee was another example that was even equipped with showers to support the women in getting used in practicing personal hygiene. Rijksarchief Overijssel, inv. 435.1, Zwolle, Netherlands.

gardening at local schools. One teacher reported that she herself lifted out strips of stones in a schoolyard to start a garden after being ignored in previous requests for a proper piece of land. Some schools even managed to keep chickens in schoolyards, but practical learning of new ways of caring for other farm animals was difficult to tackle. Only Catholic boarding schools established farms on their compounds, where pupils farmed as part of their studies and chores, while also providing food for the school. There, the teachers combined explanations of theory with their instructions for training in practice. Milking, dairy, and poultry—traditionally tasks for women—received special attention in separate courses in some regions. Other matters were not addressed at all. Baking bread, for example, which was gradually moving from the farms to commercial bakeries, received no encouragement from teachers because not having to bake bread lessened women's workload considerably. Preserving traditional food heritage for later generations seemed to be no reason for teaching.

For some subjects, teachers could only illustrate their lessons with printed media, such as posters hung up in the schoolrooms or displayed during traveling courses. The posters could illustrate plant varieties and livestock breeds and also the effects of diseases. Some teachers brought microscopes to the classroom to show microbes and the effects of deficiencies of vitamins to underline the relevance of learning about health, food safety, and appropriate cooking for each family member, whether babies, young children, sick, or old. Farm circumstances also required addressing the need to secure and use clean water, modern sanitation, and waste removal to ensure healthy food processing and preparation.

Changing the diet proved to be extremely difficult. Men especially complained that they really needed their fat and beans. Some husbands and fathers refused to eat new ingredients, such as tomatoes, potatoes boiled in their skin, or uncooked vegetables in a hotchpotch. One student remembered: "You had no encouragement, yes of my mother, but not my father. He wanted the same old food, especially big beans with bacon and meat and a lot of gravy, lots of fat, and no vegetables."[34]

Food, Health and Livelihoods in Economic and Wartime Crises

The economic depression of the 1930s brought new worries. In 1934, inspector Greta Smit notified farm household management teachers that the original ideas for farm household management education were no longer appropriate for meeting the changed circumstances. Indicative for her was the steadily decreasing number of farm household management pupils among the smallholders. According to Smit, the goal of raising the standard of living for their farms and households and stimulating new needs should be dropped; instead, attitudes needed to be changed.[35]

Smit lobbied hard for a way to better reach out to these women as well as to reach female land laborers. The new Act for Vocational Education did not provide for education or extension to adults, and overall budget cuts did not help her cause either. In 1935, however, with the support of the Ministry of Education, agricultural organizations, and health organizations, Smit managed to establish a private foundation to support rural household management extension (Stichting voor Huishoudelijke Voorlichting ten Plattelande, or HVP). Smit became director while retaining her position as state inspector for farm household management education. The newly founded HVP continuously monitored the deteriorating conditions of farm and other rural families. In 1936, HVP reported alarming malnutrition among rural land laborer and smallholding families. Mothers, according to these reports, were omitting the more expensive protein-rich ingredients from family meals so they could sell them. Taking governmental crisis support was the last thing mothers wished to do. So the teachers recommended that women keep at least a goat for milk and a pig for meat. One report read: "Because there is no home slaughtering anymore, and no money for buying extra food, the food has become more and more limited. One may be eligible for so-called cheap food products such as canned meat and meat loaf, artificial butter specially labeled as 'crisis' margarine, etc., but that was felt as the last recourse."[36]

One of the farm household management teachers reported that most of the rural laborer and smallholding women had never heard of farm household management education. They would profit enormously from any elementary knowledge, in particular on nutrition and on how available materials and products could be used more effectively. She wrote: "Often, I found out that they did not know how to bake an egg, make a simple batter, gravy, a tasty hotchpoth, or cook chocolate milk, etc. Of cooking times they had not the slightest idea. For instance, red cabbage had to be put on the fire in the morning to be eaten at noon. It was very useful as well to tell the women about nutrition. For them it was new territory, and it often attracted a lot of interest."[37]

Besides organizing individual advice, HVP established a series of six small courses to reach more women. Each course was six weeks, one afternoon a week for several hours. The courses could be taken independently of one another and even the one-penny fee could be waived. Farm household management teachers in service of the HVP taught three courses on nutrition. The first addressed summer nutrition, including canning, and the second took up winter nutrition, including slaughter. The third nutrition course addressed the specific dietary requirements for babies and for elderly, weak, and sick persons. Two other courses, one for children and one for adults, taught by needlework teachers, dealt with clothing and especially addressed how to best use all material, how to make new

Farm women in canning class, a nutrition course in a pub offered by the Dutch Foundation for Rural Household Management Extension (HVP) during World War II. During the German occupation, this formally private foundation remained the only functional institution to support farm women until the end of the war. Its courses, available to rural women of all ages, were instrumental to guaranteeing food security by teaching about nutrition, the use of wild food sources, and substitutes for scarce ingredients. Women learned to preserve produce for winter use by canning and other methods. Atria, Institute on Gender Equality and Women's History, Inv. IIAV-603, Amsterdam, Netherlands.

clothes out of old ones or waste materials, and how to do the laundry. A horticultural teacher taught the sixth course on how to produce vegetables and fruits more effectively.

The curriculum had to be limited to the basics, with demonstrations, but in nutrition courses the teachers usually cooked enough food to taste what was prepared or even share a small meal together at the end of the lesson. Again, classes had to be improvised in various buildings in the villages. According to one teacher, classes were "taught in farm barns and farm kitchens, in attics, in shops, abandoned schools, warehouses, church vestry rooms, municipal council rooms, the mayor's room, pubs."[38] Course members were asked to bring their own paraffin burners, pots, and also ingredients to be used. As one teacher notes, this provision did not always work out spotlessly: "Once, one of the women brought a bottle of home-made berry juice we would use in a recipe that day. We had put all we needed on the lawn outside because a meeting was being held in the room we otherwise used. While explaining, I tried to open the bottle. But even before the cork was pulled out, it popped out with a loud bang and the costly juice was spread all over the attentive faces of the pupils. From red to blue and lurid purple, all cheeks and hair were colored instantly."[39]

In 1938, the extension foundation also provided support and secured supervision for the newly established Agency for Smallholding Farm Families (Dienst Kleine Boerenbedrijven, or DKB). The DKB appointed eleven farm household management teachers, one in each province, who coordinated courses, gave individual advice, and when possible even lobbied for financial or in-kind support.[40] To get acquainted, these teachers organized farm parlor afternoons (*praatmiddagen*): "Then, the farm household management teacher demonstrated something on cooking, and I, the teacher in needlework, took a closer look at the old clothes that were brought with them."[41]

All these related activities of education and extension were managed and coordinated by farm household management teachers. With the support of agricultural and health authorities, they advertised their courses in the press, on radio programs, and at market stands. One wartime radio script from 1941, for instance, pointed out the relevance of the courses by reporting that they would also teach how to use inexpensive substitutes and wild foods, especially field and dune plants and various kinds of berries.[42]

Sometimes the HVP was called the little sister of farm household education.[43] Evidently the gradually expanding farm women's organizations and the farm household management teachers at schools helped organize the courses and identify women that needed support. The teachers proved to be welcome resources, responding to "questions . . . on many issues: concerning the laundry, how to convert or mend clothes, how to repair toys, how to prepare simple sweets, etc."[44]

The expanding network embracing farm and other rural women's development certainly helped the women overcome the tough economic depression and the war that was to follow. It supported agricultural modernization and rural development from the rural and farm women's perspective. Women were not only newly encouraged to economize through efficient use of resources but also to recycle available resources from the farm, garden, and "old" materials. They were encouraged to cooperate with each other, improve their self-esteem, and develop a feeling of equality with other population groups.

Starting with Sunday Modernization?

The farm household management teachers were, of course, in quite a different position from their pupils. They were the so-called experts, but as unmarried, often young women, with better-off backgrounds, they found themselves often challenged to defend their point of view within all the organizations, institutions, and situations in which they participated. Because of their social background, the women teachers had easy access to the males in charge because they were part of their family and class relations. Together, they became important spearheads of reform within the male-dominated agricultural professions, and thus the first to stand up for farm women's needs.

Teachers needed a pioneer spirit for the job. In their daily work, they had to resolve many questions and conflicts immediately. Most of the time, the teachers had to arrange all kinds of practicalities before starting with a course, such as the availability of a room, water, furniture, and equipment. They had to explain again and again what was needed and for what reason, yet because of the voluntary character of this education, pupils could easily refuse to come any longer. The teachers would try to clarify signs of distress and unusual behavior they did not understand. They would also ask the women directly which classes they found useful for their work at home. Sometimes the course periods were changed, holidays were delayed or advanced, subjects of the curriculum extended, added, or skipped, and extra lessons, excursions, or lectures were arranged by mutual agreement. But the women's wishes could not always be fulfilled. For instance, lessons in needlework were eagerly requested but seldom offered within the part-time courses because of the already loaded curriculum and the lack of teachers qualified to teach the skill in a rural setting.

The reports of the farm household teachers were usually optimistic but also indicated an awareness that they had to gain their students' interest and establish authority. A farm household management teacher reported in 1915 from Groningen that during her home visits many pupils showed that they had applied what was taught.[45] Her colleague from the same province wrote that the parents in Veendam showed much interest during her house visits.[46] Five years later, from a region in Overijssel with more smallholdings than in Groningen, another teacher was more cautious, saying: "Much has been improved, but there is still much we would have desired to be otherwise. Maybe this will change over time, especially since the farm woman has shown that she is open to making changes and admits that she feels well under this change."[47]

For instance, women cooked the new recipes at home and used fresh meat more often, but mostly they used the usual home-salted or smoked meat.[48] Farm women developed their own strategies for introducing new recipes and avoiding complaints. They presented new recipes first as something special, for instance on a Sunday or a holiday.[49] When the teachers found out, they considered it quite understandable that not all their advice could be applied to the daily routine immediately. Happy with the changes, one teacher stated after her house visits in a smallholding region: "Where the women still have to devote a lot of time to the farm, it is just not possible to spend more time on the preparation of meals, etc., . . . and indeed, the interior should stay simple to save time. Still, it is good to see that changes are achieved, for instance, by introducing stoves into simple kitchens that used an open fire."[50] Farm women also gradually adopted new methods for food conservation. In the Netherlands, as in many other countries, vacuum sterilization or canning became very popular as a new method to better preserve nutrients. Farm women were used to conserving food for the wintertime but could not always afford the new glass canning jars or time for experimentation. Some accomplished the changeover gradually. In addition to salted meat and vegetables, they initially stored some food in glass jars, then increased the amount every year. Teachers helped by providing less costly alternatives, such as reusing ordinary glass jars and using cheese and laundry cauldrons for sterilizing, which most farm women already had at their disposal.[51] The visible results of canning were very reassuring: one of the pupils of Winterswijk felt nothing could harm them when looking at all those glass jars full of preserved food in the cellar.[52]

Changing farm interiors also required a careful approach. Changes could not be effected without approval of both the farm man and woman in charge. In some regions, grandparents had the last word. Replacing the open fire with a stove, the bed box with a bedstead, and small windows for bigger ones, as well as decorating rooms with light colors and removing full-length curtains proved to be more acceptable when neighbors or family members supported the idea or had already made the same changes themselves. The arrangement of the kitchen furniture and equipment was a woman's own territory, yet the teachers observed that younger women had to deal with the opinions of the older generation. Usually, when the grandmother was living on the farm, she was in charge of the kitchen and the cooking.[53] One strategy to stimulate discussion was to take the women and their families to a farmhouse whose renovation had been done and financially supported by a development program, with the understanding that the family would open their homes for tours.

Teachers often referred to a variety in circumstances when trying to understand why pupils did not immediately apply what they had learned. One account from 1920 cites the lack of money as decisive, based on the teacher's experience that farm women in better circumstances or who performed a decreasing amount of agricultural labor had more control of farm household decisions.[54] A teacher for parts of Gelderland points to lack of schooling and prior practice.[55] Another gives a similar report about Limburg: "Sometimes, it is too difficult for them, because they left school as early as possible, and then they were kept home for working, both in the household as on the farm. They hardly have any knowledge about household subjects. Most have hardly or never cooked, which changes little by little where the woman in charge has given over a bit of her control."[56] Compromise turned out to be the usual way of dealing with conflicting opinions. According to the teachers' reports, such compromises were familiar and also practiced among themselves. These conciliating attitudes slowed changes among women who might have been entitled to demand a greater share in agricultural innovation and rural development from our contemporary perspective. The new farm and rural women in their various roles did not generate spectacular changes. Still, with innovation beginning in a cautious way, if only on Sundays, without radical signs of campaigns or resistance, a silent revolution had been started that would have immense repercussions in the long run.

Conclusion

Vocational education for farm and rural women and their daughters became an integral part of agricultural modernization and rural development programs. The prospects of the late nineteenth century, in which agriculture seemed to promise professional positions to women, had dwindled by the first part of the twentieth century. Women who pursued such professional careers became rather marginalized or invisible. They could be found in health-oriented, self-supporting communities, in biodynamic and later organic farming or gardening, and in the agricultural sciences. These new women and their activities should be further investigated and also connected internationally to the mainstream story of food, farm women, and agricultural modernization and rural development. In this area, the rural sociology group around Heide Inhetveen at the University of Göttingen has made a good start with their research on pioneer women in agriculture.[57]

For the majority of farm women, the construction of a new professional identity rested on co-ownership or tenancy of the family farm and included their specific farm labor domains. Farm household management education reflected this in its placement within agricultural education and in its curriculum that addressed the women as farm housekeepers with extended responsibilities on the farm but as solely complementary to the male farmer. Although it was frequently said that the women had to be prepared to take over in times of illness, death, or other emergencies, there was no claim to any education for total farm management. Farming in their own right was not a real option for women, a situation that was hardly ever disputed.

Farm women's work in food production for both commercial and family use—food processing, storage, and preparation—were, however, an important part of their daily lives in most of the regions. Professionals acknowledged this role and planned education for rural women accordingly. Changes, however, such as the externalization of the dairy processing, the mechanization of women's field labor, and the shifting of pig- and poultry-keeping into a specialized farmer's domain gradually diminished the farm labor of most women. Still, the education on food for farm women continued to be strongly influenced by the farm family household in which the women had to work.

The programs developed for women in agriculture reflected important elements of empowerment for rural women. Lifelong learning, self-organization, and leadership programs for women from farm backgrounds were established, and farm household management education also fostered the expansion of specific educational networks for all rural women. These successfully facilitated a type of "sisterhood" that empowered rural women to exchange and lobby for their interests, especially in the economic crisis and wartime of the 1930s and early 1940s.

Farm household management education also reflected the diversity of farm types, sizes, and customs. It largely mirrored the existing diversity in wealth and thus, implicitly, the existing social inequality. Emphasizing a common cause and cautious attitudes helped overcome the differences. Daughters of better-off farm families and rural professionals became the teachers of the less fortunate. They were the new women bringing professionalization to the majority of farm women, for whom modernization and tradition had to be negotiated in many ways. Many combined class loyalty, through which they could profit from the positions of their brothers and fathers, with gender loyalty toward other women with farm backgrounds. Both professionals and practitioners learned from each other. They learned how to bridge the existing differences of class or wealth and region, the practices of "old" times and the "new," and how to work with the diversity of both traditional and "new" women. The challenges they faced and overcame make these women's histories both interesting and important. In this sense, agricultural innovation and rural development programs for women add a new dimension to food history.

Part 2
EXPERTS

4

"Montana Extra Selects"

Harriette Cushman's Quest to Market Montana Eggs for Montana People

AMY L. MCKINNEY

The Forgotten Hen
Montana was a mining camp
And then cattle came.
And many's those that thought our state
Would never be the same.
Though sheep and cattle filled the range.
There still was fields to take.
An' so the land was filed by those
Who thought they'd surely make
Their stake from golden grain.

But all the way the lowly hen,
She'd pay the grocery bill.
Though wheat goes up and wheat goes down
She's sure to pay it still.
They make great plans to save the wheat
With such as triple A;
But nothing's done for our old hen,
'cause hens aint built that way.
They thrive in drought or rain.

They say she's still a live stock state,
An' you can not alter nature,
An' poultry's just a little thing

With men and legislature.
But while the outfit's fussin' 'bout,
Sayin' hens aint worth a candle,
(Just cause they can't drill hen seed in
Or farm 'em from the saddle.)
Montana's poultry's on the gain.

Harriette E. Cushman included "The Forgotten Hen" in her 1935 report on poultry for the Montana extension service, reflecting what she saw as the unappreciated place of chickens and eggs in Montana agriculture.[1] The egg industry had gone through many ups and downs, and Cushman maintained the importance of poultry to the economic vitality of the state. As she wrote in the poem, even though poultry hens played an important role in the economy, they had received little attention. When she arrived in the state in 1922, she lamented that "Montana was the dumping ground for inferior chicks from the midwest."[2] Cushman set out to improve the quality of Montana eggs through education, encouraging egg marketing and cooperatives, introducing a superior quality egg that was properly graded, and initiating state law to maintain standards. Her efforts had mixed results, but she did bring about significant changes to the production and marketing of eggs in Montana.

Cushman received a degree in chemistry from Cornell in 1914 and pursued a career that would challenge herself intellectually and professionally and allow her economic independence. Although it was common for rural women to work with poultry and to often take charge of turkey, chickens, and eggs on their family farms, it was unusual for a woman to be in a position of authority within the agricultural industry. Cushman maintained an important role in the production and marketing of eggs in Montana as State Poultry Specialist from 1922 to 1955. She was the first woman poultry specialist in the United States. In fact, she was the first woman to be hired as a topic specialist in any field in the extension service besides home demonstration. Throughout her thirty-three-year career as poultry specialist, only one other American woman held the same position, Cora Cook in Minnesota.[3]

When Harriette Cushman took over as poultry specialist in Montana in the summer of 1922, she concluded that poultry work in the state suffered from two main problems: poor production and poor management. When it came to management, she believed the poor quality resulted from "the inability of the producer to put the product in a marketable condition."[4] This was especially true for eggs. At the time Montana imported more than 50 percent of its eggs because of the lack of quality and quantity of the eggs produced within the state. The issue of marketing was especially problematic given the distance from Montana to the major national markets. In addition, because of the size of the state and lack of good roads, it was difficult to even reach local markets. Cushman believed the solution was "the forming of some sort of an organization which will care for and regulate the sale of their produce."[5]

Even before Cushman's arrival, however, Montana did in fact have a history of cooperative efforts when it came to eggs. According to R. L. Smith, the previous poultry specialist, to help market eggs Montana had "egg circles"—"association[s] of several neighbors living around a shipping point who organize for the purpose of marketing eggs cooperatively."[6] Because eggs were not a huge enterprise for most farmers at the time, many did not produce enough to warrant shipping to out-of-state markets. They usually sold at local markets and got "whatever [was] offered usually in trade. By combining with neighbors and pooling their eggs a sufficient number are obtained to ship to some distant market where the price is higher."[7] Throughout the 1910s and early 1920s, eggs prices seemed high enough to warrant efforts to continue the "egg circles." By 1922, however, "the Pacific Coast eggs were starting to come into the state. With their advent prices began to flatten out."[8] But the egg circles did provide a foundation for producers to pool their efforts when it came to egg marketing.

With the increased importing of eggs into Montana, egg producers in the state needed to produce a higher volume of quality eggs. To help encourage an increase in production, Smith had also initiated an egg-laying contest in 1921 to convince more Montana farmers "that money can be made from chickens on the average ranch, under average conditions."[9] The following year, when Cushman took over as poultry specialist, she increased the efforts to improve the egg situation in Montana. She maintained that quality needed to improve to compete with the eggs coming from the coast and to create a greater demand and therefore greater profits for farmers. She pushed for a stronger education program, which she hoped would result in a greater demand for quality eggs in the state.[10] With education as her focal point, Cushman was very supportive of finding innovative ways to reach a large audience. Butte, as the location of the only egg market in the state, became her chief target. She wrote in her 1923 annual report that the Woman's Club in Butte "sent the request that they would like to undertake a plan for developing a better market for home poultry products. . . . It would be the educating of the Butte housewives to demand a *good* Montana product instead of asking for eggs from the coast henneries, which they are doing now."[11]

In response, Cushman met with key organizers for an egg show to be held in Butte. The committee included Peter Potter, exhibits chairman of the Chamber of Commerce, J. L. Dorsh, a local produce man, and Mrs. Scott Fries, a member of the Woman's Club. They needed to have a product worthy of marketing, an egg of highest quality. The committee was concerned about the egg situation in Montana and believed the show would be a good platform for educating both producers and consumers about the possibilities for eggs. The competition from Pacific coast eggs was a major concern. Cushman and the committee members were adamant that quality Montana eggs should receive as good a price as the coast eggs, but until the quality of Montana eggs improved, this could not happen. Quality would need to be standardized to give consumers confidence and produce stability for the egg market. The result was the creation of the term "Montana Extra Select," which would be the mark of the highest quality eggs in Montana. At the time, the United States did not have a standardized grading system for eggs. To receive the label of "Montana Extra Select," eggs had to be properly candled and graded. With that standard in place, consumers could be assured eggs with this label would be top quality.[12] One aspect of grading eggs was candling, which determined the interior quality of an egg.[13] By candling eggs, graders could determine the quality of the egg without having to break the shell. One of the most important aspects to consider when candling was the appearance and position of the yolk at the center of the egg.[14] Eggs labeled as "Montana Extra Select" had to meet several standards, including uniform size and shape, specified minimum weight, strong shell texture, and firm egg white.[15] As Cushman explained, "Montana Products for Montana People" (a slogan especially supported by the Butte Chamber of Commerce) did not mean having a product that was "just as good" as other states but something that was superior to the competition. In her efforts to "booster" the Montana eggs, she maintained that she was "thoroughly convinced that once patrons start purchasing 'Montana Extra Selects' they will never wish to return to indiscriminate purchasing."[16] These were the eggs the egg show committee wanted to market in Montana and highlight at the upcoming show.

The next step was to introduce and educate people about Montana's quality eggs. The first Butte Egg Show was held March 6–8, 1924, "for the purpose of giving producers free advertising and consumers a chance to become acquainted with quality eggs."[17] It drew support from twenty-five counties and had a total of 109 entries. The show, modeled after a previously successful apple show, provided demonstrations on the nutritional value of eggs, proper packaging, and candling. As a newspaper in Cascade County explained, "The purpose of the show is to demonstrate to the Butte people that a Montana egg is equal in quality to coast eggs or any other eggs produced in nearby states."[18]

One goal of the Butte Egg Show was to demonstrate clearly the differences in quality of eggs to help encourage both farmers and consumers to select only top-graded eggs. One particular display was by Clara Desjardins. Her prize-winning hen, Lady Red Metal, was prominently displayed with 269 eggs, representing the number she had laid the previous year. Alongside this was a "scrub" hen and the mere sixty-nine eggs it had produced.[19] The cooking and handling of eggs was another component of the show. Local Butte restaurants featured egg dishes throughout the week. The Montana Power Company also disseminated a booklet, "One Hundred Ways to Cook an Egg," to every housewife who attended the show. Another popular display was a brooder and incubator, timed so that the eggs would hatch during the show.[20]

Getting Montana eggs to market was a primary concern and therefore careful packaging of eggs was another main exhibit. Sponsored by the American Railway Express Company, the exhibitors demonstrated how to properly pack eggs for shipment. The display included one improperly packed case. The eggs on the top layer were "beheaded," their tops broken because of their nonuniform size or the low-quality separators. The visual difference was to help producers see the proper way to package and to demonstrate a way to prevent loss in profit.[21]

Cushman stated that the "most interesting entries were those from the county associations. They not only made a big display, but they showed the foundation of a future state association which will efficiently handle the egg crop of Montana."[22] The county exhibits became an even stronger focal point in the next two eggs shows. In 1925 the Butte Chamber of Commerce awarded a prize for the best county exhibit. The exhibits were judged on the number of individual exhibits, the quality of the eggs, ease of setting up the exhibit, and the artistic arrangement and originality of the display. Big Horn County won the highest scores and was awarded the 1925 cup.

The initial egg show was deemed a huge success, and there was immediate support for organizing one for the following year. Cushman was hopeful that the interest in the show would encourage more demand for quality eggs. She believed Butte would lead the charge because it was "the people of Butte who are intensely interested in better Montana eggs. It is up to the producers to handle their eggs better and to organize in such a manner that a larger volume of fresh eggs can come onto the Butte market. [They] expressed the feeling that Montana eggs would be more desirable than coast eggs, could the eggs be obtained of a similar excellence."[23] The egg show was held for three years and maintained large crowds and numerous displays. After the 1926 show, however, Cushman and Potter, chairman of the egg show committee, decided to forgo any future shows. They believed that "for the time being, the project had accomplished all it could." The shows would return once "some sort of a cooperative marketing machinery could be established in Butte."[24] The 1926 show was the last held in Montana under Cushman's tenure as poultry specialist.

The egg shows introduced Montanans to "Montana Extra Select" eggs, but now Cushman had to convince farmers to make the effort to produce them. Cushman believed that Montana eggs could be just as good as the eggs imported into the state, but she had to be able to convince the people of Montana. One issue she faced was that there was no consistent manner of grading and candling eggs in Montana at the time. Cushman, along with the Dairy Division of the Montana Department of Agriculture, set out to standardize the process. As Cushman explained, "the United States has no standard grades for eggs as has Canada." Even though large market centers such as New York, San Francisco, and Chicago had grades, they were not exact, and although there was "an effort on foot to establish a uniform grade for the whole United States," it was going to be a long time before that would become a reality. In the meantime, Cushman believed that it was "advisable, in order that the greatest profit may be realized from a product of the highest quality, to have an establishment of egg grades within the state of Montana."[25] There was initial success for those who participated in poultry picnics throughout the summer of 1925. People brought their eggs to be candled and graded. The eggs that met the standards were placed in "Montana Extra Select" cartons and shipped to Butte. The standardization of grading for these eggs guaranteed superi-

Harriette Cushman at the 1924 Butte Egg Show. Cushman was determined to raise the quality standard of eggs in Montana. The egg show provided participants with recipes using eggs, information on their proper handling and packaging, and how to check for and improve their quality. Montana Extension Service, "Report of the State Extension Poultry Specialist, 1924," 29; accession 00021, box 87, folder 36, Montana State University Archives, Bozeman, Montana. Courtesy of Merrill G. Burlingame Special Collections, Montana State University.

ority of quality and as a result "the growers were able to realize ten cents more than they would have locally even after paying transportation."[26]

Under Cushman's guidance and leadership, "Montana Extra Select" became the mark of superior quality eggs in the state. To be packed in "Montana Extra Select" cartons, eggs had to be inspected by a licensed grader. Only those eggs that were of superior quality would receive the "Montana Extra Select" seal issued by the Montana Department of Agriculture. Cushman designed the "Montana Extra Select" carton and submitted it to the Self-Locking Carton Company of Chicago and also sent it for a U.S. patent.[27] The carton and the seal would be proof to the Montana consumer that these eggs, from a local market, were just as good if not better than imported ones. Using the Montana Women's Clubs' magazine, *Montana Woman*, to educate consumers, Cushman urged that "home-keepers of the state should familiarize themselves with these eggs and know what the carton looks like. When buying such an egg the seal must be unbroken, otherwise you are unprotected. Remember, this is a guaranteed egg you are buying."[28] Cushman's goal was to have the "Montana Extra Select" carton and seal be all that was required to prove quality to consumers.

The grading and candling of eggs became a key component to assure the highest quality for Montana eggs. Cushman wrote to M. L. Wilson, the head of the Department of Agricultural Economics at Montana State College, of her efforts to establish grades for eggs produced in Montana. She reported that she referenced "all available sources of information, both of U.S. and Canada" in establishing grades.[29] All eggs would be graded by licensed graders using the required candler and equipment. The minimum equipment required to properly inspect and grade eggs included a proper candler apparatus and a room that could be darkened, an egg scale, a standard thirty dozen cases, "Montana Extra Select" cartons, the memo of agreement for producers, and a book or card system to make entries of eggs received.[30]

Cushman, along with Chester C. Davis, the administrator of the Agricultural Adjustment Act, drew up the requirements for a grading system in Montana. The highest level in Class I (fresh) eggs was, of course, the "Montana Extra Select" (or Grade A). This was the signature egg of Montana that Cushman wanted to become the standard for all eggs produced in the state. Grade B eggs were equal in quality to the "Montana Extra Select" except in weight. Grade C eggs could be the same weight as the "Montana Extra Select" but did not maintain the same quality when it came to firmness of the egg white and the yolk's being visible but moving freely within the egg. The lowest grade eggs, Grade D, had watery whites and heavy yolks.[31]

Some progress was made for the egg industry when in the summer of 1930 a meeting was held at the Leggat Hotel in Butte to begin discussion about the possibility of writing legislation to oversee egg grading in Montana. The two main organizers of the meeting were B. R. Thrailkill, the dairy commissioner, and C. B. Quinn, who oversaw the feed and egg department at the Missoula Mercantile Company.[32] The first attempt to present a bill to the state legislature was a disaster. As Cushman recounted, "The poor egg grading bill certainly suffered. It was pigeonholed, dragged out, killed, resuscitated, torn to pieces, revamped and squashed so many times that we all despaired of its ever becoming law."[33] Even with the difficulty in the legislature, the bill did finally pass.

Although the initial 1931 law was fairly weak, it did establish the first legal standards for egg grades in the state. It was one thing, however, to pass the law, and another to enforce it. As Cushman lamented, "most unfortunately the Legislature did not make proper provisions in the law for handling disobedience."[34] As a result, Cushman once again focused on the need for education as the best way to enforce the law. By developing the educational programs regarding egg quality, Cushman hoped "to teach producers to put out a better product, buyers to pay on the basis of grade, and consumers to develop a taste for quality eggs and begin demanding such a product."[35] Even after the egg law had been in effect for nearly a decade, Cushman reported that she continued to "struggle with the egg law and its educational aspects. But merchants have been slow to put their egg business on a graded basis. They sell on grade because the law requires it. However, they do not educate their customers or push quality eggs."[36]

Cushman devised several plans for egg education, including grading demonstrations, cooking schools, 4-H egg quality demonstrations, and leaflet information. Consumer education was a central feature of Cushman's plan to improve the egg situation in Montana. In 1940 she met with Bess McClelland, who was the state nutritional specialist at the time. Cushman explained that together they "decided that something ought to be done about consumer education in eggs. Montana had a good egg law, though compliance with the law has much to be desired. It is as with all laws, only as good as folks want it to be. But how to reach consumers was the difficulty. An egg is an egg and that's that."[37] Cushman and McClelland devised the "egg quality demonstration," which would target older 4-H Food Club members. They believed that "consumers would happily give the 4-H member twenty minutes on any regular meeting program. And once there, if the member did a good job, the consumer was bound to be exposed to egg education. How right we were—at least on the first premise."[38]

Even though the efforts were often met with resistance, Cushman's spirits were revitalized with each success. She reported that "sometimes when we despair of making any headway with grading or making retailers and consumers quality-conscious, we receive a pleasant surprise."[39] One such surprise happened in 1944 when the manager of the Safeway store in Granite County requested a meeting with Cushman to arrange training for his poultry producers. She remarked that "this was such an unusual request that the Specialist inquired into Mr. Brockway's history." He had worked with a demonstration farm cooperator who had his state egg grader's license. Cushman was happy to report that "after learning how greatly quality was appreciated, Mr. Brockway wanted to repeat the practice of selling only graded eggs."[40]

She was also inspired when she learned of better practices being used in stores. One particular area of concern was the refrigeration of eggs. She reported that "from casual observation the practice of stores to keep eggs under refrigeration has grown rapidly. There have no doubt been many times 8 retail merchants who have adopted the practice in 1953. It is becoming increasingly difficult to observe a store which is not following the practice."[41] Refrigeration was one of Cushman's priorities in consumer education. She wrote a flyer, "Know Your Eggs," in which she asked:

> Where does your grocer keep his eggs? Where do you keep yours? Heat is eggs [sic] worst enemy. Eggs should be handled in the same way that milk is. When the temperature gets above 50 degrees at any time from the nest to the frying pan, a chemical change takes place in the white. It is similar to the souring of milk. Only in an egg, the white becomes thin and watery. An egg may drop from AA to C grade in a single day if kept at room temperature. Demand that your grocer keep his eggs in the cooler and at the same temperature he keeps his milk and butter. See that you put your eggs in the refrigerator as soon as you receive them.[42]

Cushman knew the flyer was making an impact when she told of "an interesting incident" when she was "checking out of a hotel at Missoula." The clerk informed her that upon receiving one of her bulletins, he had "learned a lot." He had found it in an egg carton of Cushman's design at the grocery store. He admitted that "they were fine eggs but stated that he never knew before that any special care was needed for eggs in the home nor that eggs required low temperatures in cooking." Cushman was convinced that "incidents like these prove that consumer education can be accomplished by bulletin material."[43]

Picnic that concluded the study club meeting of the Lake County Egg Pool. One of the goals of these gatherings was to educate pool members about egg quality. Members could also bring their eggs to be candled and graded, with the possibility of earning the "Montana Extra Select" status. Montana Extension Service, "Report of the State Extension Poultry Specialist, 1935," 54; accession 00021, box 87, folder 47, Montana State University Archives, Bozeman, Montana. Courtesy of Merrill G. Burlingame Special Collections, Montana State University.

She was also encouraged to know that graders and producers were being more stringent with the eggs accepted into the state. She reported that "prior to this year it has been the custom to dump low grade mid-western eggs in this state at certain seasons. This year it began again. Two cars were distributed in Butte. Mr. Thrailkill's [chief of the Dairy Commission] deputy caught these up and had them returned to the wholesale house where they had to be recandled and properly graded. . . . As a result, for the first time in 15 years a car of low grade eggs moved through Montana and [was] dumped in Spokane."[44]

As demand increased during World War II, Cushman was concerned about how to better distribute Montana eggs. To reach more consumers and preserve the quality of the eggs, Cushman recommended a new method devised by Mr. Dissley, the manager of the Fergus County Creamery. Dissley explained in 1942 that by "breaking-out eggs and sharp freezing them," the eggs could then be "'flash candled' viewing 3 dozen in their fillers at a time." After careful inspection of the eggs for quality, removal of shells or foreign objects, the eggs were placed into tin cans and flash-frozen. Once they had a full shipment, the frozen cans of eggs were shipped to the egg drying plant in Spokane, Washington. Cushman explained that "as a result of this practice the price to the growers never dropped below 25 cents a dozen and most of the time stayed above 30 cents."[45]

Maintaining high-grade eggs during the war years was also problematic. As Cushman explained, "the average farmer has neither time nor inclination to produce or market top grade eggs. This coming season with tire and gas shortages, eggs will come into central points in even worse conditions." Cushman was concerned about the surplus of low-grade eggs in the state and maintained that the flash-freezing method Dissley had developed provided a viable solution. As she explained, "the surplus marketing group will not purchase these poor eggs. They can not be stored. Well, let's turn around and make these poor eggs acceptable to the trade. They won't make the highest quality dried eggs but if broken and sharp frozen, they can be salvaged without undue loss."[46]

Throughout the rest of Cushman's tenure as poultry specialist until her retirement in January 1955, the egg situation continued to be problematic. As during the Depression, it was difficult to get consumers and producers to abide by the egg law. But even with this difficulty, Cushman's leadership was clearly recognized by many, and she was viewed as one of the few people who could turn the egg situation around. Although Montana never became a major egg producing state and struggled to maintain egg cooperatives, she spearheaded successful educational programs, marketing strategies, and clear standards for quality eggs. She was also instrumental in getting a state egg law passed, which regulated the quality and grading of Montana eggs. Her efforts helped increase a market for "Montana Eggs for Montana People," and, as her poem about "The Forgotten Hen" emphasizes, even though the hen may not have been recognized as a significant aspect of the Montana economy, the hen indeed meant survival to many because "she'd pay the grocery bill." Cushman was able to take a commodity that was underutilized in Montana's economy and augment it to produce "Montana Extra Select," a mark of quality and excellence.

5

The Chain of Interdependency

Apples from the Orchard to the Consumer

ANNE L. MOORE

In December 1926, Lorian Jefferson wrote to former Massachusetts Agricultural College (MAC) president Kenyon Butterfield informing him about the publication of her research project in an academic journal. The article, "The Balance of Trade in Farm Products," focused on a study of the food supply in Fitchburg, Massachusetts. Butterfield had authorized Jefferson's project, and they both expected this research to be published by the college as an experiment station bulletin. When state authorities in charge of publications rejected the article, the resourceful Jefferson found another avenue for publishing her research and scholarship. Her letter exemplifies Lorian Jefferson's determination as well as the respect and loyalty she felt for Kenyon Butterfield. Despite a myriad of obstacles against her as a woman, Lorian Jefferson was dedicated to scientific and empirical research, scholarship, and teaching. Over the course of her twenty-three-year career at MAC, Jefferson was involved in outreach and research with both producers and consumers. Her writing was published in scholarly journals and in the bulletins of the Massachusetts Agricultural Experiment Station as well as in magazines and publications directed toward consumers. In addition to writing "The Balance of Trade in Farm Products," Jefferson investigated farm ownership, exploring the circumstance of how Massachusetts farmers acquired and operated their farms. Most of her research focused on the production, marketing, and distribution of apples and onions. Beyond her research, Jefferson played an active role in encouraging women to engage in agriculture and horticulture by supporting women students at the college and through her involvement with the New England division of the Women's Farm and Garden Association. This chapter investigates the experience of the New Woman as a researcher and scholar within challenging gender stratification in academia and food production. Because of expanding opportunities during the first decades of the century, women found independence in urban environments. Most of the roles offered to women, however, did not challenge the more traditional assumption that women were essentially nurturers. The place for women in the academy was limited to the discipline of home economics. While challenging the traditional limits for females in academia, the focus of Jefferson's research and liaison work at MAC focused on identifying and demystifying the chain of interdependency between the producer and the consumer.

Lorian Jefferson was hired at MAC in October 1912, when the Board of Trustees approved Butterfield's request to "employ expert service for the Division of Rural Social Science."[1] No specific hire was named in the minutes, perhaps because, despite the need for expertise, the position was classified as clerical. Like many other educated and professional women during this era, Jefferson was hired at a level below her education and experience.[2] As the only staff member at the college with the title "Expert Secretary," Lorian Jefferson stepped into a world of change and reform within both the college and the rural community. Working directly for Kenyon Butterfield, Jefferson served a leading national figure in advocating for economic, social, and education reform in rural areas. During the Progressive Era, Butterfield and other urban agrarians focused on reforms in response to late nineteenth-century changes that had moved the United States toward a totally industrialized and urbanized society. These changes left agriculture and rural life behind, because rural areas had not experienced the economic, social, and political improvements enjoyed by urban areas. Despite these improvements, Progressive Era reformers identified an array of social and political issues intensified by immigration, crime, rampant materialism, loss of religion, disintegration of the family, and class conflict.[3] Urban agrarians believed rural life to be more wholesome than life in crowded urban areas. But for this rural ideal to be realized, economic, social, and education reforms were necessary in rural areas. In addition to the social and moral concerns resulting from urban growth, reformers worried about whether American farmers could continue to produce sufficient food for the country.

Jefferson as an Urban Agrarian

Lorian Jefferson's participation in the urban agrarian movement centered on the initiatives of Kenyon Butterfield. As a member of President Roosevelt's Country Life Commission, Butterfield helped construct three recommendations focused on further investigation and examination of agricultural and country conditions: a campaign for rural progress, facilitated through local, state, and national conferences; the formation of organizations; and, finally, a federally funded outreach program centered at the land-grant colleges. The last of these recommendations focused on training agents to offer instruction and practical demonstrations in agriculture and home economics to the rural population.[4] The land-grant colleges played a key role in carrying out these changes. Even before official approval by the U.S. Congress, many land-grant leaders had, like Butterfield, initiated programs designed to achieve these objectives. Beginning in 1908, Butterfield began an outreach program offering instruction and demonstrations to the rural and agricultural community. The following year, Butterfield wrote the first draft of the legislation that, when passed in 1914 as the Smith-Lever Act, authorized financial support from the federal government for such programs.[5] Butterfield also increased the "short course" programs the college offered during the month of January and throughout the summer in order to offer education on a variety of topics to nonstudents.

In an atmosphere of increased investigation and research, Lorian Jefferson's research skills were quickly employed as Butterfield reconfigured the role and scope of the college. As a rural sociologist, Butterfield was convinced that "a college of agriculture ought to emphasize the economic and social aspects of agriculture fully as much as the production and scientific aspects."[6] The expanded scope of the college implemented the Country Life Commission recommendation of examining the conditions of agriculture and rural areas. Butterfield established a division of rural social science, which he headed. The new division included departments of rural sociology, rural education, and agricultural economics. Jefferson spent the twenty-three years of her MAC career working within the division, including as a faculty member in the Department of Agricultural Economics from 1919 until her retirement in 1935. Beginning in 1914, she served as one of the summer course instructors, lecturing on educational surveys. For several years she taught a course on the history of New England as part of the regular undergraduate curriculum.[7] But most significantly, from 1915 on she was assigned, as part of the Agricultural Economics Department, to the experiment station. As the research component of the land-grant colleges, experiment stations received federal financial support through the Hatch Act of 1887. The chair of the Department of Agricultural Economics was a fellow University of Wisconsin graduate, Alexander E. Cance. Jefferson and Cance were colleagues until Jefferson's retirement in 1935.

Lorian Jefferson had also been a graduate student in the economics department at Wisconsin, overlapping by a year with Alexander Cance in the program. Although it is not clear if Cance and Jefferson were directly acquainted, faculty did know both of them and likely recommended Jefferson to Butterfield and Cance as a skilled researcher. After a decade as a high school teacher, Jefferson enrolled in graduate school at the University of Wisconsin in 1906, graduating the following year with a master's degree in history. She then enrolled in the doctoral program in economics. Over the ensuing five years, she completed her doctoral work with John R. Commons, for whom she also worked as a research assistant under a Carnegie Foundation grant. Before moving from Madison, she was employed by the Wisconsin Board of Public Affairs. Jefferson did not take (or was not allowed to take) the required exams to graduate from the doctoral program, but she had completed a dissertation, a portion of which was later published by Commons.[8]

In coming to MAC, Lorian Jefferson transitioned from a focus on labor economics to agricultural economics. A new discipline introduced in 1909 at the University of Wisconsin and MAC, the foundations of agricultural economics had developed late in the nineteenth century through attempts to measure and validate the relationship between input costs, output prices, and farm profits. The farm depression of the 1870s to 1890s added an emphasis on market outlets and modes of organization to gain bargaining power to these foundations. Jefferson's research closely focused on the foundations of agricultural economics, using statistical data, conducting surveys and personal interviews. Jefferson's published research identified trends and patterns as well as solutions designed to overcome the identified problems and limitations.

One of Jefferson's first publications was a joint research effort with Cance on cooperatives, published in 1915 as "Farmers' Cooperative Corporations" by the Vermont Department of Agriculture.[9] This was not her first exploration of cooperatives. In 1913 she had published a summary article,

"The California Fruit Growers' Exchange," about the structure of cooperative associations in place to benefit California citrus fruit growers.[10] Both the 1913 and the 1915 articles emphasize the opportunity that cooperation among farmers offers for lowering input costs by jointly buying supplies for growing or raising their products and for achieving greater efficiency, ease, and profit by marketing their products together. Although the California growers had experienced the benefits of cooperatives for twenty years, this was a new venture for Vermont farmers. Instructions for the legal and logistical requirements of a cooperative are also provided in the article. Its most critical message emphasizes the necessity of cooperation for the individual farmer's financial survival. Cooperatives reduce input costs, raise output prices, and thereby increase profits. The article also refutes the myth of the independent farmer, pointing out that the farmer "buys much of his food, most of his clothing and nearly all of his farm equipment."[11] The farmer is therefore part of a chain of interdependency. This notion of the producer also being a consumer offers a common ground for producers to understand the expectations of the consumers of their products. According to the cooperative article, the key to receiving the best price for one's goods is the need for products to be of known quality and grade, to be uniformly packed, with large enough shipments to meet demand, and for regularly shipping throughout the season of demand or throughout the year. The consumers' expectations for quality goods did not vary, whether the consumer was a farmer or an urban housewife.

Research on Apple Production and Marketing

The theme of interdependency and cooperation resonates throughout Jefferson's published research. After her promotion to faculty status in 1919, Jefferson conducted and published seven major studies. This research, published as both experiment station bulletins and in academic journals, reflected the Agricultural Economics Department's focus on agricultural products that could be successfully produced in Massachusetts. In a 1922 memo to Butterfield, Cance recommended that the department focus on the production, distribution, and marketing of poultry and eggs, fruits and vegetables, and dairy products. Between 1923 and 1929 Jefferson conducted five studies focused on market outlets for Massachusetts apples and onions. A consistent theme throughout these publications was consumer demands and expectations.

Although Massachusetts had not been a major producer of apples, its growing conditions suited apples, and producers had the advantage of large markets nearby. This meant that Massachusetts growers would have lower shipping costs than the competition from Washington and Wisconsin. Jefferson's studies on apples ranged from examinations of costs for the producer to tracking a shipment of apples from Massachusetts to New York City. A study published in 1925 examined the costs of marketing the 1923 apple crop, focusing on the input costs beyond the actual growing of the apple to determine the cost of getting the picked apple into the hands of the consumer.[12] The source of information for this study was Jefferson's personal visits to growers in four areas of the state. The price of apples obtained by growers varied from one to four dollars for a barrel of ungraded apples to a range of $2.75 to $5.00 for graded apples. Included in the study were the practices of a cooperative with set prices for sorting, grading, and packing, and a commission merchant who sold all of the association's apples. The total cost was estimated to be $1.80, considerably lower than that of the individual growers, who had estimated costs at between $2.00 and $3.50 per barrel. These results reinforced the advantages of cooperatives with their greater efficiency. The study also recommended marketing apples beyond the harvest season to obtain higher prices. The cooperative was able to achieve this through greater volume and access to storage facilities.

Jefferson's 1927 bulletin, "The Market Outlet for Massachusetts Apples," analyzes factors that lowered demand and suggests solutions to increase market demand. The study draws on a variety of sources and techniques, including data from the U.S. Department of Agriculture, interviews with wholesalers and retailers, and data from questionnaires. The questionnaires were distributed to consumers by mail, to school students by teachers, and also to women's clubs.[13] The McIntosh apple had become the most popular apple for eating. The greatest demand was for apples of grade A quality and uniform size, with little demand for lower quality or ungraded apples. Jefferson noted that this demand required improved orchard practices to increase the proportion of highest grade apples as well as creating an outlet for lower grade apples. How apples were purchased had also changed. Apples were now bought individually and by the pound rather than by the barrel, a change that intensified the demand for higher quality. Based on the success of advertising for California citrus fruits and Cape Cod cranberries, Jefferson recommended a systematic advertising campaign to increase apple consumption.[14]

The commercial apple crop in Massachusetts increased 65 percent between 1916 and 1925. And while the two largest markets for Massachusetts apples were New York City and Boston, markets within the state of Massachusetts were increasing as the demand for eating varieties of apples continued. Jefferson's next study examined the McIntosh apple in one of the premier market outlets. "The McIntosh Apple on the New York Market," published in May 1928, traced a shipment of McIntosh apples from the Nashoba Apple Packing Association, a cooperative in eastern Massachusetts, to New York City in late March 1928. After the shipment of apples was sold at auction, Jefferson interviewed wholesalers, brokers, and jobbers involved in handling and distributing the apples to retail outlets and the consumer.[15] Purchased by eight jobbers and one fruit store chain, most of the apples in retail stores were sold by the pound. McIntosh apples were estimated to account for 35 to 40 percent of the apple sales in New York City. Jefferson concluded that the demand for McIntosh apples in New York City was not being fulfilled; this was a response to critics who voiced concern over the number of McIntosh trees planted in Massachusetts in anticipation of increased demand. While about 35 percent of the apple trees in commercial orchards in Massachusetts were not yet bearing fruit, Jefferson's analysis of the New York market argued there was demand for future increases in production. Jefferson's earlier publications on apples had emphasized the need for better orchard practices to produce higher quality apples. This study confirmed that this goal had been met. And because the apples had been marketed through a cooperative in the eastern portion of the state, the study also supported evidence of the importance of cooperatives. A comparison of prices obtained for this shipment of apples confirmed that the quality of the Massachusetts McIntosh was equal to any received in New York City from other areas, such as New York and Washington.[16] The publication confirmed the ability of Massachusetts producers to meet market and consumer expectations. It also confirmed the benefit of the cooperatives as beneficial to the producer in reducing costs and to the market for quality produce that met consumer expectations. As the 1915 study had argued, cooperatives offered producers many advantages, including lower costs for sorting, grading, and packing, as well as the capacity to market products at times of greater demand in the market. The shipment of McIntosh apples was sold at auction in New York City in late March, long after the end of the growing season.

Beyond verifying the important role cooperatives could play, Jefferson's studies provided insight for producers into the complexity of marketing goods and how much the opinion and expectations of the consumer mattered. The 1928 study of the New York City market for McIntosh apples tracked how apples went from a Boston area cooperative to an auction in New York City and then through several jobbers and wholesalers to reach consumers at retail outlets. Its detailed explanation of how the product reached the consumer, along with the personal interviews with the jobbers and wholesalers, created a more powerful message than Jefferson's words alone. It offered producers a glimpse into the process of marketing and distributing their produce. The interviews with the wholesalers and jobbers showed the preference for McIntosh apples among the largely Jewish population of the Washington Heights area of New York City. With her account of the popularity of this eating apple in one ethnic group, Jefferson demonstrated the critical role that a product's reputation played in creating increased demand and supporting higher prices for the product in the market. New York City wholesalers and jobbers had come to regard Massachusetts McIntosh apples as high quality and recommended the apples in their homes and neighborhoods. In establishing this important connection between the Jewish wholesalers and jobbers and the success of the McIntosh apple sales, Jefferson grounded the importance of a reputation for high quality in individual experiences, offering insights that statistics on prices and sales could not have done alone and providing producers with an inside view of the expectations of the consumer.

Both scholars within the discipline of agricultural economics and academics at MAC supported the notion in the 1920s that if "a premium for certain qualities and types of products is more than large enough to pay the increased cost of growing a superior product, the individual can and will adapt his production and marketing policies to market demand."[17] Jefferson's published research on apples reflects this mantra as many of the Massachusetts apple producers adjusted to growing market demand through a variety of initiatives. They planted trees to produce the new variety, improved the quality of the produce to a higher grade, and formed cooperatives to facilitate getting a high-quality product to the consumer efficiently. Although not all growers in Massachusetts had made these changes, a large group of growers using a cooperative provided Jefferson with a persuasive example to demonstrate to producers the opportunity for success.

Onions were another important crop in Massachusetts. The Connecticut River Valley was one of the principal onion growing areas of the

United States during this period, with around forty-five hundred acres in onion production. In 1926 the experiment station published a bulletin about experimental work on onions conducted by four MAC departments: agricultural economics, agronomy, botany, and entomology. Jefferson wrote the section on the markets for onions. Her summary focuses on the readily available markets in Boston and New York City. However, foreign imports of sweet onions had recently captured a greater share of these markets. Reaffirming the importance and expectations for a uniform and high-quality product, Jefferson recommended ways to improve the quality of onions, such as immediately removing the harvest from the field rather than leaving it to dry and thus making it more susceptible to rot and spoilage, and shipping onions to market in crates rather than bags.[18]

Jefferson's Research beyond Apples and Onions

In addition to her research on the commodities of apples and onions, two of Jefferson's other major research projects took up the important question of farm ownership. Her first solo research investigation, which she began in 1919, resulted in "Farm Ownership in Massachusetts," published in 1923, both as an experiment station bulletin and in the *Journal of Farm Economics*.[19] The farm ownership study involved a survey of 650 farm owners in Massachusetts. Jefferson focused on potential input costs by identifying how the farm had been acquired, who had acquired it, how it was paid for, sources of loans, income, and the value of land and farm property. The study revealed that four-fifths of the farms were purchased from someone other than a relative or through inheriting it from a family member or receiving the farm as a gift. At the same time, the value of land had increased approximately 21 percent in each decade since 1900, reaching more than seventy-three dollars per acre during this time. Close to 90 percent of those who had purchased farms had earned and saved the money; among them a slight majority had earned the money in a business other than farming. As potential farmers saved money to purchase more expensive land, the average age of the Massachusetts farmer also increased. Comparing the results of the Massachusetts study with studies conducted in Minnesota and Kansas, Jefferson determined that Massachusetts farmers were older and less likely to have experience as laborers or tenants on farms. With the increasing cost of land, those who had acquired the farms since 1900 were more likely to have debts; more than 60 percent had loans. Banks had become the greatest source of loans as the older practice of borrowing from relatives became less common. The results of Jefferson's study reflected the increasing financial pressure on Massachusetts farmers and the need for reliable income sources. Despite the increased cost of acquiring and maintaining farms, the study found that 69 percent had reported a profit; among the profitable farms, more than half used the surplus income for improvements on the farm. Massachusetts farmers, despite the challenge of high input costs, also had the advantage of nearby urban markets.

During the 1920s several studies were conducted in urban areas of Massachusetts to track the sources of food consumed within the state. Jefferson conducted one such study on the food supply for Fitchburg. Located in north central Massachusetts, Fitchburg was a factory town with three rail lines. The railway freight records provided an accurate account of the incoming food products, in addition to which Jefferson contacted local residents, a source that led her to some important conclusions about the local market. In particular, none of the cereal or canned goods consumed were produced within the state, while all the milk and cream consumed was produced in Massachusetts or nearby New Hampshire. About half the poultry was "local," but only about 7 percent of the meat was produced locally. In total, about 30 percent of the food consumed was locally produced—mainly dairy, poultry, eggs, and seasonal fruits and vegetables.[20] Jefferson's direct contact with the residents allowed her to link consumer purchasing patterns and ethnic preferences. The Fitchburg study outlines the demographics: more than 75 percent of the residents were first- or second-generation foreign-born.[21] According to Jefferson, there were "striking differences" in the diets of the large populations of Italian, Finnish, Polish, and Greek immigrants and also the Jewish population. The range of vegetables included in the typical diet varied, as did the quantity and type of meat consumed. She found, for example, that locally produced meat was likely kosher, an important consideration for the Jewish community. She also reflected on working women and their efforts to provide midday meals for their families. Women in the Greek community, most of whom worked in factories, prepared meals quickly using forequarter cuts of lamb, which was also an inexpensive cut. Lamb, she noted, was known in Fitchburg as "Greek meat."[22]

Bridging the Rural-Urban Divide

Published in 1918, Jefferson's "The Community Market" was her only MAC Extension Service publication. With a structure similar to the Vermont cooperative publication, the bulletin articulates the advantages of community markets for both producer and consumer, emphasizing proximity of the market, direct contact between producer and consumer, and the reduced costs for both parties. The study itemizes the procedure for establishing a market, lists the necessary facilities, and identifies the roles for producers and market organizers. The article includes an appendix of documents, with sample sales sheets and vendor permits.[23]

Community market in Quincy, Massachusetts, c. 1917. A major emphasis of Jefferson's research and outreach at Massachusetts Agricultural College focused on the interdependency of the producer and consumer. The producer needed to understand consumer expectations while offering quality products as efficiently as possible. The consumer expected high-quality products at competitive prices. The community market offered the opportunity for direct interaction and savings for both producers and consumers. Courtesy of the Department of Special Collections and University Archives, W. E. B. Du Bois Library, University of Massachusetts Amherst.

Jefferson's bulletin on community markets was part of a wider campaign by the college and the state to counter the high cost of food within Massachusetts. The state population of about four million was estimated to be almost 95 percent urban, with only about two hundred thousand people living on farms. The urban population was continuing to increase, with a growth of 17 percent between 1910 and 1920. Residents of Massachusetts had, by 1920, experienced high food prices. The federal Bureau of Labor Statistics estimated that a family with an income below $150 per month would spend half their income on food.[24] Facing some of the highest food prices in the country, the Massachusetts legislature passed legislation in 1915 requiring all towns and cities with populations of more than ten thousand citizens to set up public markets. The college focused on the production of eggs, fruits, vegetables, and dairy products within the state, and local markets provided the ideal venture for producers to market these commodities directly to consumers. When the United States entered World War I in 1917, the stresses on food production and availability only increased. With rising prices and the need to provide food for troops, authorities urged consumers to practice food efficiency.

Jefferson was not officially assigned to the extension service, but she was involved in a number of the college's outreach activities to the community. The female extension service agents, who reached residents of the state through demonstrations and instruction, were Jefferson's first female professional colleagues at MAC. She collaborated with them on outreach to women as consumers. Extension agents, for example, assisted Jefferson in distributing copies of her research surveys. The service also offered direct access to rural producers and consumers. While extension work was typically divided along gender lines, with men as producers and women as managers of the home, Jefferson challenged those notions.

In her study of farm ownership, Jefferson had identified women not just as producers but also as farm owners. Her survey revealed that 3 percent of Massachusetts farms were owned by women, a total of eighteen farms. Ten of those women had purchased their farms, seven had inherited the land, and only one had received the property as a gift. Jefferson also noted the size of the farms, which ranged from eleven to two hundred acres. Five of the female-owned farms were more than a hundred acres.[25] Jefferson's findings served to correct the idea that women were solely consumers. As farmers and landowners, women were clearly agricultural producers in their own right.

Just as Lorian Jefferson was beginning her MAC career in 1912, Alexander Cance published research on rural demographics in Massachusetts. A key demographic feature, Cance found, was "the preponderance of males in small rural communities." The number of males exceeded females by 5 to 12 percent in sixty-two western Massachusetts towns. Women, Cance noted, found very few attractive vocational opportunities in rural communities. As increasing numbers of women obtained college degrees or opted for employment as clerks and office staff, women accounted for more than 35 percent of professional workers.[26] These jobs were in urban areas, not rural ones. Cance's study details the effects of a century of outward movement from rural areas within the state and emphasizes the importance of reorganizing agriculture, with the application of scientific knowledge and methods, improved transportation facilities, and better rural living conditions. On a positive note, Cance observes that some reorganization of agriculture had taken place in towns close to larger urban areas. Market gardening, greenhouse vegetables, floriculture, and small fruits had replaced staple crops. Further, Cance observed two "back to the land" movements, including in "picturesque" areas, with good highways and trolley lines, making country homes accessible and appealing to urban dwellers. The second movement was taking place in the many small towns in Massachusetts that had industrial facilities, including woolen mills, button factories, and wood product businesses that offered employment to residents. In those communities, "a host of wage earners and small salaried men and women have been looking for liberty on three acres. Some have wisely combined agriculture with their regular industrial occupations and are meeting with more or less success."[27] Waged workers, in other words, were supplementing their incomes by producing food on small acreages.

Women Bridge the Rural-Urban Divide

Lorian Jefferson embodied the tie between urban and rural experiences. Aware of the draw of urban areas for women of her generation, she was an example of a woman who had lived in a rural area and who remained committed to bridging the urban-rural divide. The late nineteenth and early twentieth century offered a variety of opportunities for women beyond the tradition roles of daughter, then wife, then mother. Like Jefferson, many women from middle-class families enrolled in college and obtained college degrees. After graduation these women joined other young women in pursuing employment in the growing array of office-based vocations, retail opportunities, and professions deemed suitable for women, such as social work and nursing. Jefferson, after graduating from college, returned to rural Wisconsin and lived with her family for almost a decade, working as a teacher. She enrolled in library school at the University of Illinois in 1900 but dropped out to return home. Two years later, Jefferson moved to Washington State to teach school. In 1905 Jefferson moved to Madison, Wisconsin, and enrolled in graduate school. Jefferson and her cohort, unlike their mothers' generation, tended to be college educated, live apart from family, support themselves financially, and participate on their own in social and recreational activities. These women also tended to marry later in life and to have fewer children. The "new women," as these independent women came to be known, found their place in urban communities.[28] But as Jefferson's own life illustrates, new women could also be found in rural and agricultural circles.

During her career at MAC, Jefferson would participate in numerous efforts to make rural life attractive to women. The solution, in the vision of President Butterfield, was education and opportunity. He envisioned a women's college of agriculture and rural life, stating in the college's 1914 annual report that "there is a new demand upon the part of young women for participating with men in the training for the agricultural vocations."[29] The college for women, however, would not be realized, and in 1918 Butterfield revised plans for educating women in agriculture by establishing a women's program within the two-year diploma program of the Stockbridge School of Agriculture. A women's placement officer was hired to arrange a six-month placement for students during their first year of the program, in addition to supporting employment opportunities after graduation. The placement officer, Margaret Hamlin, was a 1905 graduate of Smith College who had taken up farming. The first women graduated from the MAC Stockbridge School of Agriculture in 1921. The eleven graduates had majored in animal husbandry, horticulture, general agriculture, vegetable gardening, and floriculture. Hamlin's "Pioneering in Placement for Women in Agriculture," written in 1958, reflects her knowledge of nine of the eleven women after their graduation. Many worked on country "estate" farms or state correction institution farms. Of the four listed as having married, all had married a farmer or nurseryman, were living on farms, or were active in farm and garden club work.

Between 1921 and 1948, Hamlin estimated, 259 women received agricultural and horticultural placement training positions.[30]

Hamlin and Jefferson were close colleagues because of their liaison work until Jefferson's retirement in 1935. Just before Hamlin's arrival as placement officer at MAC, the New England division of the Woman's National Farm and Garden Club (WNFGA) became an integral part of the efforts to involve women in agriculture and offered an outreach opportunity for the college. Jefferson became involved in the New England division of the WNFGA in 1917, shortly after Butterfield spoke at one of the division's first conferences. She quickly became an active member of the executive board as well as committees on education and agricultural information, serving until her retirement and death in 1935.[31] The New England division formed in 1917 after the national organization's annual conference in Boston. In 1913 Elizabeth Leyton Lee, a landscape architect in the Philadelphia area, organized a conference dedicated to discussing opportunities for women in agriculture and horticulture at the Ambler School of Horticulture. This gathering led to the creation of the WNFGA. Over the next few years, the mission of the club evolved to focus on bringing together women interested in and or engaged in horticulture and agriculture, connecting the producer with the consumer, and offering opportunities for exchanging ideas and information.[32]

One of the women Jefferson would work closely with for the almost two decades of her involvement in the WNFGA was Charlotte Barrell Ware. The niece of Justin Morrill, the senator from Vermont who sponsored the land-grant college act that bears his name, Ware had been a pioneer in the "clean milk" movement early in the twentieth century. She developed the Warelands Dairy in Norfolk, Massachusetts, produced the first certified milk in New England, and developed a school to train dairy workers.[33] Earlier, Ware had been an active member of the Women's Educational and Industrial Union (WEIU) of Boston. Founded in Boston in 1877 to respond to social problems in urban areas, the WEIU's programs included a store where women could sell crafts and foodstuffs, employment training, and a retail training program. The WNFGA modeled itself on the WEIU and the Intercollegiate Bureau of Occupations, later the Bureau of Vocational Information.[34] Unlike its predecessors, however, the WNFGA's mission focused on both rural and urban reforms. Its first priority was to encourage and support rural and urban women's interdependence by bringing together women in both the producer and the consumer roles, which was first successfully achieved with a "holiday sales shop" in December 1917. Forty vendors made approximately a thousand dollars in sales, and the holiday sale became a regular event held in the Boston Horticultural Hall, evolving into a regular sales shop known as "The Green Door," which operated in Boston for more than fifty years.

In addition to her committee involvement, Jefferson became a liaison between MAC and WNFGA's New England division, a role that began when the WNFGA joined forces with the Women's Land Army of America (WLAA) to mobilize women in agricultural production as the United States entered World War I in 1917. The New England division designed training, coordinated workers, and contributed to the cost of training and placing "farmerettes."[35] According to the MAC student newspaper, Jefferson traveled throughout New England during the summer of 1918, issuing reports on WLAA efforts, while overseeing projects and assignments of female students from gardening to poultry.[36] On January 3, 1918, Jefferson was one of five speakers providing accounts of the WLAA's efforts in New England; the other speakers were the coordinators of its programs at women's colleges.

After the war ended, a more formal connection was established between the college and the club, with Jefferson serving as liaison. Continuing the wartime efforts of training, scholarships, and job placement, the division sponsored five scholarships each year. Between 1919 and 1935, the WNFGA division aided ninety female MAC students, who majored in agriculture, horticulture, and 4–H-related home economics, by providing money for tuition, fees, room, board, and books through an annual donation of six hundred dollars from the Massachusetts Society for Promoting Agriculture. Providing scholarships to enable women who lacked the financial means to attend college became a common practice among women's clubs.

When Butterfield established an advisory council for women at MAC in 1921, another dimension was added to Jefferson's liaison. While the head of the home economics department, Edna Skinner, served as chair, Jefferson was appointed as secretary and served in that role until her retirement. According to Butterfield, Jefferson's connections to women beyond the campus made her presence critical to the success of the council.[37] The council's membership included women representing each county from throughout the state, organizations such as the Grange, and different religious denominations. The council provided a mentoring group for the female students at MAC, an advisory group to the college president, and a lobbying group to the legislature for women's programs at the college. The WNFGA New England division's closer relationship and direct involvement with MAC began in October 1920, when the college organized a conference on women in agriculture in conjunction with the dedication of the first women's dormitory at the college. Speakers included WNFGA president, Louisa King, who delivered the keynote address, "Vocations for Women in Agriculture," and C. J. Galpin, economist in charge of rural life studies for the USDA, whose speech was titled "The Transition in Rural Life from the Family Regime to the Community Regime." The program also included two women sharing their "personal experiences in agriculture." The first speaker, Nettie Caroline Burleigh, had graduated in 1898 from Mount Holyoke College. After several years of teaching, she established Wonheim Farm in Waterville, Maine. Burleigh's talk at the 1920 conference, with its claim that "farms need cultured women," captures the efforts of Jefferson, Butterfield, and the New England division of the WNFGA.[38] Burleigh, who exemplified the New Woman, advocated that women like herself should become involved in agriculture. She spoke with authority based on her personal experience. While the success of the Women's Land Army during World War I had proven women's ability in agriculture, Burleigh provided an example of this success over a longer span of time.

According to a 1912 article published in *The Maine Farmer,* Burleigh and her companion, Miss St. James, had begun farming in 1909 with a small flock of quality laying hens. Within two years of the first shipment of eggs, the farm was shipping fifty-six dozen eggs weekly, which yielded more than a hundred dollars per shipment. The farm raised high-quality poultry stock and hatching eggs, and Burleigh also successfully experimented with various crops, including wheat, oats, and corn. By 1924, when Burleigh wrote an article for the *Bulletin,* Wonheim Farm had a letterhead boasting "eight-rowed flint seed corn, fresh butter, eggs, poultry, home-cured pork and bacon, all shipped by mail or express." Burleigh had been elected as a selectman in Vassalboro, Maine, and listed her activities beyond farming as town business, politics, and lecturing on farming.[39] It is not surprising that Jefferson asked Burleigh to write about her experience as testimony of the potential for women in farming for the 1924 special edition of the *Bulletin.* The importance of the collaboration between the New England division of the WNFGA and MAC to train—

and create opportunities for—women in agriculture and horticulture is conveyed in Burleigh's conclusion: "I believe there is no greater need today than the need of efficient men and women, of refinement and education and business ability, on our farms."[40]

Conclusion

The theme of interdependency between producers and consumers resonates throughout Lorian Jefferson's research and outreach work. Beginning with her joint research with Cance on cooperatives, Jefferson established that farm enterprises were not independent entities. The producer, she asserted, is also a consumer aspiring for the highest quality food, equipment, and clothing. To reduce costs related to getting the product to market, producers needed to form cooperatives. The producer-consumer interdependence continues as the consumer makes choices about products such as apples and onions. Producers are also dependent on banks for loans to assist in the purchase and operation of their farms. The consumer, as Jefferson's research points out, depends on the producer sending high-quality products to market at the lowest possible price. Jefferson's work with the WNFGA also exemplifies this chain of interdependency. Solutions such as community markets and the Green Door offered direct transactions, highlighting the interdependency among women as both producers and consumers.

Lorian Jefferson was a typical New Woman. The male members of her family held professional vocations as ministers and doctors. After attending college and then teaching high school for a decade, Jefferson returned to college, seeking a different career. Strongly independent, she remained single throughout her life, had an active social life, and traveled extensively, both for work and leisure. As a woman, she experienced discrimination and loss of opportunity during her career. Although she completed the requirements of the doctoral program at the University of Wisconsin, she was not allowed to take the exam required to gain the degree. Although she offered "expert service" at MAC, her job title relegated her to clerical status. Nevertheless, as an educated woman who pursued advanced education at a land-grant college, she proved herself to be a dedicated and resourceful scholar as well as a skilled researcher. Over the course of her twenty-three years at the Massachusetts Agricultural College, she and Alexander Cance formed the backbone of the Department of Agricultural Economics. Because of the high turnover of faculty and researchers, Jefferson served as department head during Cance's sabbaticals and absences, all the while teaching, conducting research, and doing outreach beyond the campus. Quiet and demur, Jefferson seldom appeared in faculty photographs of the period. As one of the first female faculty members at the college, Jefferson was a model for other female faculty and students, yet her contributions have gone largely unrecognized by historians.

6

Women's Institutes in Canada and the United Kingdom

The Weighty Matters of Domestic Science, Home Economics, and Food Security

LINDA M. AMBROSE

On June 4, 2015, Queen Elizabeth tried, without success, to make the ceremonial first cut in a hefty fruitcake that had been prepared to mark the centenary of the Women's Institute (WI) in the United Kingdom.[1] The cake defied that royal effort, adding a bit of levity to the historic occasion, while also serving as a metaphor and reminder that this rural women's group, despite its attention to tea parties, cakes, and jam, should not be taken lightly.[2] Indeed, as this chapter argues, the role of food in the history of rural women's organizations should not be trivialized; it provides weighty connections to the science of public health, the economics of agricultural production, and the assurance of food security both private and public. Food studies have long been at the heart of groups such as the WI and their efforts to educate women and encourage them to take more prominent roles in rural economic and political life. The tendency in popular culture has been to dismiss these women's gatherings as nothing more than inconsequential "tea parties." Among historians, Nancy Tomes makes the point that "when the 'second wave' of feminists began to criticize the ways scientific experts had fostered restrictive roles for women, home economists' teachings on the subject made an easy target."[3] By extension, such groups came to be associated with that same conservative and limiting emphasis on domesticity. It is time to revisit conclusions that assume the concerns of women's organizations were trivial matters and reconsider how the WI's attention to food was more substantive than stereotypes have suggested.[4]

Adelaide Hoodless became a recognized expert in domestic science after her young son died from drinking impure milk in 1889. One of the founders of the Women's Institutes in Ontario in 1897, she is also significant for her 1898 textbook, *Public School Domestic Science*, which was used throughout the province of Ontario. Her family tragedy motivated her engagement with food science, food security, and education. XR1 MS A001, Adelaide Hoodless Family Papers. Courtesy of Archival Collections, University of Guelph Library.

The centenary cake in June 2015 marked the anniversary of the founding of Women's Institutes in the United Kingdom, but the WI was actually a Canadian institution that traced its origins back to the community of Stoney Creek, Ontario. Local women formed the first branch in 1897 with the aim of bringing women from their community together and providing them with training in home economics and encouraging their involvement in local initiatives, especially concerning public health. The origin of the WI is well known among Canadian rural women, arising out of one Ontario woman's tragedy and her subsequent activism in promoting food safety. Adelaide Hoodless, who is regarded as the founder of that first WI in Stoney Creek, became an advocate for teaching women the science of homemaking after her fourteen-month-old son, John Harold Hoodless, died in 1889 from "summer complaint," a gastrointestinal illness caused by drinking impure milk. Hoodless was heartbroken. As the wife of a prosperous businessman from Hamilton, Ontario, and an experienced mother of four children, she was shocked that this kind of personal tragedy should touch her life. In response, she became a tireless champion for the cause of educating women about how to care for their families by applying a scientific approach to homemaking.

In 1898 Hoodless wrote *Public School Domestic Science,* a textbook that was widely used in schools throughout Ontario. Given the nature of her son's death, it is not surprising that the book placed great emphasis on handling milk in home kitchens, declaring that "all milk should be sterilized or pasteurized in warm weather, especially for children."[5] In fact, Hoodless's book reflected the science of the day and the recent emphasis on germ theory. "Milk absorbs germs from the air and from unclean vessels very readily," she wrote. "It should never be left in a sick room or near water pipes. Absolute cleanliness is necessary for the preservation of milk; vessels in which it is to be kept must be thoroughly scalded with boiling water, not merely washed out with warm water."[6] Hoodless had no academic credentials, but she wrote out of her pragmatic conviction that this kind of information was desperately needed among Ontario homemakers. Historians influenced by the second-wave women's movement have regarded Hoodless's intervention into domestic science with suspicion because it seemed to smack of elitist tendencies to impose unrealistic middle-class standards on the population at large.[7] If one looks at such initiatives through the lens of food security and food science, however, there is no doubt that Hoodless's text engages with the science of the day and that it provides basic science information for future homemakers who otherwise would not have had access to the information at that time.[8] Rather than providing grounds for dismissing Hoodless's efforts and the rise of domestic science simply as mechanisms to "keep women in their place," food history and rural women's history invite a reconsideration of the complex relationship between domestic skill, expertise, and feminism.[9]

Although Hoodless may not have been a proponent of women's suffrage, she is celebrated as one of the founders of Women's Institutes in Canada, an organization that grew out of concerns over the need to apply public health measures to the home. Emphasizing domestic science education for rural women, providing opportunities for women to socialize and overcome rural isolation, and encouraging women to engage in community activism, the WI grew in popularity among Ontario women, particularly after the provincial department of agriculture began funding it in the 1910s. The local WI clubs regularly offered educational opportunities to women in their communities through a series of short courses and monthly meetings that included peer instruction about best practices in food safety and domestic science.[10]

Twelve years after the first WI was established in Ontario, similar groups were created in British Columbia. While improving public health and extending domestic science education would also be important in British Columbia, its model was different from Ontario's. By bringing the WI to the rural communities of Canada's westernmost province, B.C. women hoped to use the WI as a means of encouraging women to earn a living from their agricultural work and of prompting them to enter the political arena. This difference in mission can be attributed to the influence of Madge Robertson Watt (1868–1948), the same woman who would later found the WI in the United Kingdom.

Madge Robertson Watt got her first taste of country living when her husband accepted a government appointment as the medical inspector at the quarantine station on Vancouver Island in 1897. It was in that idyllic setting that Watt began to cast herself as rural, although her elevated social status, postgraduate education, and published popular writings meant that she was hardly living the life of a typical Canadian farm woman. Although Watt lived in the countryside when she began her work with the WI, her background set her apart from most of the women the organization was meant to help.[11] In 1909 she turned her attention to the establishment and expansion of the Women's Institutes, using her class privilege to influence government initiatives for rural women across the province. The ideas Watt transplanted from her life as an Ontario university student, journalist, and clubwoman found receptive soil among policy makers looking for strategies about how best to meet the needs of rural women in British Columbia.[12]

Although Watt's previous experiences were mostly urban ones, when she helped establish her local branch of the WI and began working with provincial authorities to promote the movement, she billed herself as an expert on rural women. The historian Jo Burr Margadant asserts that historical actors represent themselves (or "perform") in particular ways depending on their changing circumstances and settings.[13] The concept of performance is a useful way to explore Watt's adoption of the persona of an expert on rural women because, given her previous experiences, it is almost as though she was playing at being rural, adopting a posture that would prove most effective in her efforts to influence government policy.[14]

Watt was a highly educated woman, and her cultural authority stemmed in part from her wide experience as a writer and journalist. Given her new social location as a well-to-do woman in rural British Columbia, it is not surprising that she stepped into a leadership role.

When Watt turned her attention and her influence to rural women, she sometimes lamented the food habits she observed in rural British Columbia. For example, she reported that she was sorry to see rural B.C. households consuming commercially canned goods rather than producing their own food.[15] That observation, from a woman who enjoyed rural living without having to worry about making a living through agriculture, certainly illustrates a rather idyllic notion of rural life. And yet, a closer examination reveals that Watt's proposals for rural women went well beyond nostalgia or romantic ideals; indeed, they included elements of the New Woman outlook, with emphasis on expanding the roles that women could respectably occupy. As a woman with professional qualifications herself, Watt was part of a growing number of thinkers who believed that women had the right to meaningful paid work, higher education, and involvement in political life.[16] These ideas about women, especially prevalent among British and American writers of fiction at the turn of the twentieth century, were still fiercely debated in Canada at the time. When Watt forged her new identity as a rural woman, she did not compromise on her commitment to expand female horizons, including those of women who lived outside the city and who did not enjoy the same privileges that were afforded to her because of her elite class status. Watt proposed real economic alternatives that would allow rural women to make a viable living in the countryside. This was quite different from the usual "country life" discourse that called for the modernization of rural homemaking so as to relieve rural women from hard work. It was also different from the agrarian view that emphasized the injustice of "unequal exchange between farmers and merchants," though Watt did agree with agrarianists on the point that women could use cooperative marketing schemes to their advantage. But the heart of what Watt called for was entrepreneurial agricultural training for women that would allow them to generate a sustainable income and join the growing ranks of urban women who had found gainful employment. In effect, Watt imported her urban ideals about new womanhood into her recommendations for improving rural life in British Columbia.

When Madge's husband, Alfred Watt, accepted the position as inspector of quarantine, he uprooted his young family from their comfortable urban life in Victoria and relocated them about eighteen miles away, to William Head, a rural setting on a peninsula near Metchosin. Because there was no road from their home to Victoria, their travel was always by boat.[17] The archival records reveal some important details about the Watts' new life. The young doctor's new job paid very well; when he was appointed in 1897, he was earning $2,500, commanding a salary that is well over $300,000 in today's terms.[18] Although the Watts lived on rural acreage near the quarantine station, complete with some prize sheep and ample space for gardens, they could well afford to include hired help in their household.[19] One of Madge Watt's new hobbies was gardening, an arcadian pursuit that many of her class took up at this time. According to David Demeritt, arcadianism was an expression of rural living that "enjoyed . . . widest currency among the urban middle classes," with its emphasis "more on leisure than livelihood."[20] Madge Watt was particularly renowned for her cultivation of roses, but she also grew some of her own food.[21] In addition, the Watts claimed multiple honors for her horticultural skill and food preservation at the local fall fair in 1910, when Madge won prizes for her loganberries, gooseberries, and raspberries, and her jams and bottled fruits, while Alfred won prizes for the fruit from his trees, including pears, peaches, and nectarines.[22]

Clearly, the Watts embraced country life. Madge's affinity for urban living intersected with her newfound enjoyment of the rural experience as she strategized about how she might influence other women in rural communities across the province. Her food hobbies gave Watt more than personal satisfaction and prestige in the eyes of the fall fair judges. Based on her own rural experience, Watt came up with ideas about what other rural women might accomplish as a way of making a living in the countryside. She found her background of education and working for women's advancement to be quite compatible with her new rural setting. When she asked herself how she might combine all the elements of her life and make a difference in the lives of other women, she found her answer in the Women's Institute.

The Women's Institute first came to British Columbia in 1909, when the provincial government, in cooperation with the B.C. Farmers' Institutes, invited Laura Rose to address groups of women about her area of expertise, dairying. Rose, born in 1866 in Ontario, had lived with her bachelor brother on a farm in North Dakota, where she performed typical domes-

Laura Rose, the "Lady Instructor of Farm Dairy" at the Ontario Agricultural College, traveled widely demonstrating dairying techniques and was instrumental in the creation of the Women's Institutes in British Columbia, Canada. RE3 OAC A137. Courtesy of Archival Collections, University of Guelph Library.

tic tasks, including managing the poultry and producing butter, tending the garden, and keeping house. When her brother married, Rose returned to Ontario, where she attended a ladies' college and later enrolled at the Ontario Agricultural College in Guelph as a student of dairying. Historian Margaret Kechnie describes Rose's career path this way: "After completing her studies with distinction, she was hired by the Ontario Department of Agriculture to work in its travelling dairies project. In 1898 Rose was the first woman appointed as an instructor in dairying at the OAC, and she taught there for thirteen years. Throughout her career Rose was in great demand as a speaker and as a judge at rural fairs. Her articles appeared in numerous farm journals in both Canada and the United States."[23] In 1899 the Ontario government appointed Rose to serve as an organizer for the province's Women's Institutes, a role that led to an invitation to visit British Columbia. There is no doubt that Madge Watt looked to Rose as a potential model when she considered how Women's Institutes might be transplanted from Ontario to British Columbia.

In addition to her work at the Ontario Agricultural College and with the Ontario WI, Rose also wrote a textbook, *Farm Dairying*, which was published in 1911.[24] The book enjoyed wide circulation as the standard text for dairy schools working with local farmers. Rose was accustomed to giving talks and demonstrations among Ontario farming communities, and her book was geared to small producers, offering instructions on how to ensure the quality and safety of their milk by addressing hygiene, health, and affordable applications of equipment that met the standards of scientific agriculture. Indeed, Rose's work as an expert in dairying relates closely to issues of food security.

According to Kechnie, Rose was very traditional in her views about gender. Rose's instructions lead Kechnie to conclude that "Rose felt women's only role in the dairy was to keep the milking utensils clean and the male milkers' clothing spotless."[25] It is tempting to dismiss Rose's emphasis on cleanliness and sanitation in the dairy barn as a sign of her commitment to essentialist gender roles. A closer look at the science of dairying that she promoted—particularly her insistence on standards of cleanliness as the mark of best practices—makes her commitment to food security clear. Given Adelaide Hoodless's domestic tragedy twenty years earlier, securing high standards in home dairying practice makes sense. Not only was Rose insistent on women's qualifications for this work, she was also giving women the access and knowledge they needed to protect themselves and their families.

If Kechnie is correct that Rose was comparatively conservative, then Rose and Watt were hardly kindred spirits. Still, Rose was considered an expert in home dairying and in organizing women's groups. Thus, Watt recognized that she could be a useful ally in spreading Women's Institutes to Canada's most western province. When Watt hosted Rose in British Columbia, her dairying lectures and organizing efforts met with great success, and the first branches of the WI were established on Vancouver Island that fall.[26] Among the communities that welcomed the WI was the one in Metchosin, close to Watt's home near the William Head quarantine station. The organizational meeting of the Metchosin WI was held on October 4, 1909, and Watt was elected one of the directors of the branch.[27] Based on the model from Ontario, where WIs had been meeting since 1897, monthly meetings of local WI branches became avenues of education, sociability, and activism for rural women.[28] It is a model that Watt strongly endorsed, seeing in it the potential to improve the lives of rural women.

There are, however, significant differences between the WI branches in the two provinces. In Ontario, the movement got off to a slow start,[29] but it caught on quickly in British Columbia: within one year, there were twenty-one branches in the province, with a membership of 590 women. Funding at the rate of fifty cents per member was provided by the provincial Department of Agriculture with assistance from the Canadian government; after 1913 those funds came through the Agricultural Instruction Act.[30] These grants were created with the goal of enhancing rural life for "better and happier men and women."[31] Over the next few years, the B.C. Women's Institutes grew steadily, from twenty-five branches and 974 members in 1911 to forty-seven branches and 2,857 members by the end of 1914.[32]

Beyond the rate of expansion, another significant difference between the WI in Ontario and in B.C. is tied to different visions of the roles that rural women should play on the farm. Looking beyond the conservatism that Kechnie describes as typical of the Ontario WI, Watt nevertheless took elements from that model. She was convinced that the WI would be an effective organization among rural women in British Columbia. Leveraging her convictions about women's advancement, Watt used the services of Laura Rose, the Ontario dairying expert, to garner interest in rural clubs for B.C. women focusing on education about food issues and women's relationship to food. For Watt, women and food were inextricably tied together; women were not only consumers but also agricultural producers.

Food production and female producers were on Watt's mind in 1910, when she had the opportunity to speak to the Royal Commission on Industrial Training and Technical Education about the situation of rural women on Vancouver Island.[33] The Canadian government had charged the commission with gathering information about the state of industrial education and technical training across the country and determining what help the provinces needed in implementing or improving vocational training programs. One speaker, representing Victoria's Local Council of Women (to which Watt also belonged), argued that there were few vocational training opportunities for women in Victoria and that urban women working for pay had limited opportunities for advancement. The council was reminding the commission that because "the skilled worker is one of the nation's most valued assets, this training would be not only an individual, but a national benefit."[34]

When it was her turn to speak, Watt called on the government to expand opportunities for rural women as well. She was convinced that work as producers of food could lead to sustainable incomes for women. Therefore, she recommended that the commission should find ways to help rural women profit from work such as beekeeping and herb production and that it should create short courses for women on topics such as dairying and poultry raising that would help them earn an income. She argued that such courses would benefit not just the women who already lived in rural British Columbia but also the immigrant women streaming in from countries such as India, China, Japan, and Australia. Because Watt would eventually go on to work among international women's communities, it is noteworthy that this is the first time she demonstrated a concern for women of different ethnicities and nationalities. Watt advised the commission that the government of Canada could best help women by strengthening its support for the Women's Institutes, with their great potential to effect change.[35]

In many ways, Watt's appearance before the 1910 commission marked her debut as a self-declared expert on rural women. And though the recommendations she made are consistent with her New Woman ideals, some of her language is further evidence of the arcadian notions she held. According to the commission's report, Watt waxed eloquent about rural living, stating that "the ideal farm life is a beautiful one,—the soothing sounds of the country and the twittering birds, the children engaged in happy pastimes outside, the housewife filling the larder with apples for the winter, the home life around the country fires."[36] She tempered this romantic portrayal of farm life, however, by voicing her concern that "the farm home is too often a house where they are living on [commercially] canned foods. It is hoped that the Commission will remedy that condition."[37]

For Watt, the Women's Institutes were more than simply another good cause to which she could lend her name or a poetic way to express her enthusiasm for the idyllic rural life she was enjoying. Watt did not share Rose's views about limiting women to their so-called womanly sphere. In the WI, Watt saw a vehicle that she could use to promote her vision of expanding women's roles in work, education, and politics. Seeing the WI's potential to help rural women improve the quality of their lives, she became so committed to the movement that she accepted an appointment to the British Columbia WI advisory board in 1910. Watt devoted herself to this new role, bringing to it her penchant for argument, her knack for effective expression, and her boundless energy.[38] She had devoted her earlier career to expanding possibilities for urban women; now she turned her attention to expanding rural women's prospects through their food work.

At the board's first meeting in August 1911, held in the chambers of the provincial legislature in Victoria, W. E. Scott, deputy minister of agriculture and superintendent of the B.C. Women's Institutes, explained the reasons for forming the board, including the need for "expert advice on affairs coming under his supervision which related especially to women."[39] The board would assist the provincial government in strategizing about how best to provide domestic science education, health care, and various other kinds of support to rural women in their roles as homemakers and mothers. Though the function of the board was an advisory one, he added, "unless the suggestions should be contrary to the policy of the Government, the recommendations of the Board would be carried out by the Department."[40] Watt was one of four women on the board, each representing a different region of the province. At that first meeting, Watt was chosen to serve as secretary of the advisory board, a role she occupied for the next two years.[41]

That small group of four women had a great deal of power over the agenda of the province's WI movement. Not only would they be advising the government, they would also be meeting annually to set and review priorities for the WI. This included determining the types of domestic science instruction they should provide. At their first meeting, Watt and her colleagues maintained their focus on food when they determined that a lecturer on cooking should be engaged and that young married women especially needed instruction in a range of domestic topics.[42] Board members would also travel throughout their districts promoting the WI and helping women create WI branches in their own areas. Perhaps it was on the strength of her own experience living in the country over the preceding thirteen years that Watt agreed to serve in this capacity and to cast herself as an expert on the needs of rural women, but in her suggestions one can also see how her years of city living shaped her recommendations. Indeed, three New Woman ideals are evident in the work that Watt undertook as an advisor to the B.C. government about what women in the province needed: access to paid work, opportunities for continuing education, and influence in political matters.

Watt's first priority—creating greater opportunities for rural women to derive an income from their work and encouraging them to do so by establishing their own WI branches—was a significant departure from the priorities of the Ontario Women's Institutes during this period and thus perhaps the clearest indicator of the influence of New Woman thinking on her work with the WI. Kechnie's study of the WI's early years in Ontario (1897–1919) argues that "WI leadership intended to use domestic science to draw women closer to the farm home and to deprive farm women of their income-producing work."[43] Highly critical of the Ontario WI leadership's agenda to impose domestic science education on farm women, Kechnie argues that "had women embraced the philosophy of domestic science as articulated by Adelaide Hoodless and other WI leaders, it would have narrowed, not increased, women's sphere of influence and would have made women, particularly married farm women, even more economically dependent on the men in their lives than they already were."[44]

When Watt suggested forming Women's Institutes for rural B.C. women, it is clear that the model she had in mind diverged significantly from Kechnie's description of the Ontario model. During that first meeting of the advisory board in August 1911, Watt recommended that the B.C. Department of Agriculture send experts to speak to rural women about how their work with food, specifically gardening and poultry keeping, could be lucrative. Visiting lecturers should be encouraged to make "special reference to the commercial aspects for women."[45] This emphasis on paid work is reminiscent of ideas Watt had promoted in the 1890s when, before her marriage, she had published a serial column called "By the Fireside" in the *Toronto Globe*. As a journalist, Watt repeatedly made a strong case for why women should pursue meaningful paid work and offered strategies for how they might best accomplish that goal.[46] Such progressive ideas, which appealed to her urban readers in Toronto, still informed her proposals when she turned her attention to the needs of rural women in British Columbia and their earning potential as agricultural producers.

It is also possible that the entrepreneurial spirit of the women she was living among in British Columbia influenced Watt to emphasize potential schemes for generating income. Through detailed exploration of wealth accumulation and entrepreneurial activity among urban women during the late nineteenth and early twentieth centuries, historian Peter Baskerville has found that women in Victoria were quite active in the local economy as landowners, credit takers, and businesswomen as compared to their peers in urban communities such as Hamilton, Ontario.[47] Scholars believe that this greater economic activity among women can be attributed to British Columbia's frontier setting,[48] which could have been another reason that Watt encouraged the entrepreneurial spirit among the rural women with whom she worked. Whatever may have motivated her, it is certain that Watt had encouraged women to seek economic opportunities for themselves long before her move to British Columbia. She was convinced that rural B.C. women could earn an income if they embraced their potential as agricultural producers.

A second priority for Watt was providing educational opportunities for women. Again, this was a shared concern of the "new women," and from the beginning of her involvement with the WI, Watt constantly emphasized that rural women would enjoy and benefit from books and literature. According to the minutes of the very first advisory board meeting, Watt "spoke of sending a dramatic speaker, who would give selections of good literature, and accompany the lecturer sent out from the Department. The members agreed that this would be a source of information and enjoyment, and it was resolved to take this up at some future occasion."[49] It is quite likely that Watt pictured herself as a candidate for the role of dramatic speaker about literature; certainly, it was the kind of talk she would have been perfectly comfortable delivering, given her university training and her penchant for public speaking. In Watt's view, good reading material was critical for women in rural settings, and not simply for enjoyment; in addition to good literature, she encouraged women to read educational materials, especially domestic science. Unlike the leaders of the WI elsewhere, however, Watt wanted women to read this material not simply so that they could retreat into what Ontario experts like Hoodless and Rose called their "womanly sphere," but rather to learn about strategies for making a profit from their agricultural work. Watt recognized that the settler society in British Columbia was different from the older, more established, and more conservative rural population of southern Ontario and that revenues generated by women's work as food producers could be significant.

In addition to the related goals of paid work and suitable reading materials for their continuing education, Watt emphasized a third element in her WI work: women should be informed about and involved in the political process. The kind of political influence that she envisioned can be seen as an example of what historians call maternal feminism, in which women extended their mothering traits beyond the home to their local communities and increased their presence on the public stage. In 1910 Watt gave a talk titled "Women as Citizens" to a rural community north of Victoria that was planning to create a local council of women. According to the local press, "Mrs. Watt spoke of the uplifting influence which could be exerted by women in the ruling of the schools and beautifying the school grounds, in keeping the public streets clean and fresh, in serving as school trustees and in setting up a high moral standard for children and young people in amusements as well as in work."[50] Landscaping the schoolyard would be only the first step toward these women's greater involvement in public affairs, and while Watt urged them to begin with improvements to their local schools, she expected that they would also exert their influence in the affairs of the school board and other forms of local government.

In a format mirroring that of a WI meeting, Watt's talk was followed by a lecture on food science, specifically, how to preserve and sterilize milk and "the most approved methods of keeping it pure and fresh." Again, this illustrates how, for Watt, organizational life, activism, and food were closely entwined.[51] Watt would have felt completely comfortable with the format, which combined a talk that promoted the extension of women's political influence with something as practical as health concerns over pure milk; the program offered precisely the kind of content she promoted for the WI.

In the spring of 1911, Watt presented a talk titled "Women's Institutes and Women's Suffrage" to the Victoria branch of the B.C. Political Equality League.[52] Her title suggests that she saw potential in the government-sponsored Women's Institutes as a vehicle for promoting greater political involvement for rural women.[53] In recognition of her tireless efforts on behalf of the Women's Institutes and her many other causes, Watt was one of two women appointed to the senate of the new University of British Columbia in 1912.[54] In October that year, after being named as a delegate by the WI advisory board, Watt also represented British Columbia at the 1912 International Dry-Farming Congress in Lethbridge, Alberta. These concurrent appointments, to the university senate and the farm congress, further indicate that Madge Watt was moving in influential circles. Her fulfilling such prestigious roles on behalf of the provincial government points clearly to her status as a recognized expert in rural affairs. These commitments also illustrate her continued interest in both rural and urban causes and the close tie between the history of rural women's movements and food history.

After she left British Columbia for England in 1913, Watt continued her work on behalf of rural women on the international stage. Living in rural Sussex, she was instrumental in introducing the WI to England and Wales in 1915. Watt did that wartime service in part as an act of loyalty to the Empire during World War I. In light of the very real threat of a German naval blockade, she believed that members of local Women's Institutes might help to solve the problem of wartime food security.[55] Watt successfully exported the made-in-Canada idea of Women's Institutes to the mother country in what has been called an act of "reverse colonialism."[56] For her wartime service, the British government made her a member of the Order of the British Empire.[57] Watt's career as a professional advocate for rural women was part of a longer history of women's stepping into roles that combined expertise in rural life and food issues with activism for and by women's groups. As a transnational figure, Watt illustrates the ties and networks that existed between experts on rural women in Canada and the United Kingdom.

After World War I, the WI remained strong, although Watt worried about its becoming "chiefly social." Instead, the WI in the United Kingdom came to bear the imprint of Watt's influence.[58] The WI in England and Wales was built on a model similar to that of the B.C. Women's Institutes, with its focus on helping women generate income from their work in the countryside. Perhaps the greatest evidence that this emphasis was important in the British groups is the network of WI markets in which WI members could sell their surplus goods. By the 1920s the markets were successfully operating on a cooperative basis. As WI historian Maggie Andrews explains, "WI markets ranged, then as now, from being open one day a week, to every day, and sold on commission. They did not restrict themselves to WI members and often sold work produced by smallholders, the disabled, and ex-servicemen."[59] The WI markets still operate today, renamed as "country markets," and are renowned for their success in establishing significant incomes for women.

In 1933 Watt took her work to the next level when she became the founding president of the Associated Country Women of the World (ACWW). The history of Watt's role in founding that group, which has been told elsewhere, is an important part of her legacy. The organization she helped found currently draws its membership from more than seventy different countries, continues to represent the concerns of rural women on the international stage, and holds consultative status with the United Nations.[60] Women's agricultural production work is a central theme for the ACWW, and food security remains high on the organization's agenda. In 2015–16 the group lent its support to several global campaigns, including World Food Day and Sustainable Development Goals where the "ACWW recently signed up to the Ban Ki-Moon's Zero Hunger Challenge. Our pledge obligates us to not only work to eliminate hunger, but also to double smallholder incomes, in a resilient and sustainable way."[61] ACWW has also lent its support to World Soils Day because, as their website explains, "we fail to connect soil with our food, water, climate, biodiversity, and life. We must invert this tendency and take up some preserving and restoring actions."[62] Most recently, the ACWW has joined forces with the UN's Food and Agriculture Organization to promote 2016 as "International Year of the Pulses." In this effort, the goal is to "heighten public awareness of the nutritional benefits of pulses as part of sustainable food production towards food security and nutrition."[63] Surely these kinds of projects, combining the science of the day with concern for the economic well-being of women in agriculture and global concern for food security, closely resemble their founders' priorities.

When the Queen attempted to cut the cake at the centenary celebration in London in 2015, she had to pause to ask for assistance with the task. When scholars of food history attempt to slice through the past, they too may need to pause and appreciate that there is great substance to what women such as Adelaide Hoodless, Laura Rose, and Madge Watt accomplished in rural women's clubs. From arguing for the science behind pure milk to earning money from their agricultural production and working to ensure wartime food security, rural women in groups such as the WI have a long and complex transnational history of engagement with food.

Part 3

EXTENSION

7

The Indefatigable Mrs. Webb

Food, Radio, and Rural Women—A Legacy of World War I

MAGGIE ANDREWS

On Tuesday, April 2, 1935, the indefatigable Mabel Elizabeth Edwards, who was known to the British Broadcasting Company only as Mrs. Arthur Webb, launched her new radio series, *Farmhouse Cookery*. The program listing in *Radio Times*, BBC's weekly magazine, introduced the series and Webb in the following terms:

> Hidden away in the farmhouses of Britain are wonderful recipes, and in spite of "progress," traditional dishes still exist if they can only be found. And so that most popular of cookery broadcasters, Mrs. Arthur Webb, who has sold close on 160,000 copies of her talks in pamphlet form, is to go round the farms of Britain on a voyage of discovery. Junkets in Devon, saffron cake in Cornwall, tripe in Cheshire, Eccles cakes in Cumberland, pork and onion dumplings in Norfolk and Suffolk, she will track down the farms in which they are made and wheedle the farmer's wife into giving her the recipe.
>
> Mrs. Webb is herself a farmer's daughter, worked for the Board of Agriculture during the war, and now writes for the *Dairy Farmer*. She has been reconnoitering Yorkshire this last week, and this morning she will come to the microphone and describe a typical farmwife of the country, something of the country around and the farmhouse itself—to say nothing of the dish she serves up for great occasions.[1]

In *Farmhouse Cookery* and the book that accompanied the series, Mrs. Webb provided instruction on how to cook a variety of regional rural food, knowledge she had acquired from the farmers' wives in Britain thanks to a tour she had recently undertaken at the BBC's expense. The series was both the pinnacle and the epitome of Mrs. Webb's career as a professional rural woman, a classic example of how this determined women straddled a number of supposedly contradictory categories. She was both rural and suburban, both an expert and merely the conduit for the expertise of other rural housewives. She seemed to suggest that women's domestic skills were not scientific but natural and innate, yet she earned money teaching these supposedly "natural" domestic skills to other women. She performed a very traditional gender role while undermining the assumptions about women's natural gender role, and thus caused "gender trouble."[2]

Mrs. Webb's success was in many ways a consequence of particular cultural, economic, and social pressures and influences. Her idealized version of rural domesticity appealed to the zeitgeist of the moment. The legacy of World War I meant that both food and the rural home had moved to a more culturally central place, at the heart of the nation.[3] Interwar Britain was experiencing unemployment and economic uncertainty. Anxieties and concerns were frequently expressed about the dire state of British cooking and the demise of the supposedly feminine domestic skills. Webb's advice on regional rural food became popular in part because her recipes were not "expensive or difficult"; rather, they enabled housewives "to produce attractive dishes without the expenditure of too much time, labour and fuel."[4] The recipes were ideal for the busy housewife who listened to advice on the radio.

During World War I, radio developed into a system of communication that was able to convey military information, but the BBC did not begin broadcasting to the general public in Britain until November 1922, and radios did not become a mundane item in most British homes until the 1930s.[5] Radio was a domestic medium, consumed in the home, and focused on the home. In time it came to be a signifier of the home, as the *Radio Times* suggested: "In these days wherever there is a home there is a radio set. Also wherever there is a radio set there is something very reminiscent of a home."[6] In interwar Britain the domestic sphere was strongly associated with women, who were quickly identified as a key radio audience. Commentators suggested women were "the best listeners—the most receptive, the most attentive, the most appreciative and in the best sense the most critical."[7] This was important both within the BBC and beyond. Broadcasting was quickly seen as having great potential as an educator. One working-class, lifelong member of the Labour Party described radio as "providing a university education for those without the benefit of the 1944 Education Act,"[8] a measure that provided secondary education for all in the years after World War II.[9]

The educational and social value of radio was seen to be important for rural women, particularly as an antidote to the loneliness and isolation of their lives. In the late 1930s, *Radio Times* observed: "We often hear of the blessings that radio has brought the lonely countryman. . . . Here it is that radio is so invaluable, for it provides other voices, which drive out the melancholy ghosts with laughter and music."[10] Alun Howkins has argued that in the interwar years "few operations of the agriculture cycle functioned without the work of women,"[11] yet not all women relished living in an isolated farmhouse. The London *Times* noted that "one of the few farmers who admit to making money was heard to say . . . that his livelihood had been threatened because his wife refused to stay another winter in the country" and then went on to suggest that he found "the solution to the problem" by buying a radio.[12] Similarly, Mrs. Mackay, who lived in the wilds of Colleden Moor, explained to the *Radio Times*: "You cannot understand what the wireless means to us up here."[13] Radio, it was hoped, would broaden rural women's horizons, make them think, and educate them. "Talks" were seen as a key in this, as the BBC *Yearbook* explained in 1928: "Whether the talker appeals to his listeners by illuminating family things, or by lighting up strange and remote subjects, he may set moving trains of thought, sparks of imagination, a wish to read more, to see more, to learn more and perhaps to do more."[14]

Throughout the 1920s the BBC experimented with a range of programming for housewives, who were seen "as the feminine monitors of domestic life."[15] The programs developed into *Household Talks*, a regular morning slot on weekdays between 10:45 and 11:00 A.M., and a wide range of programs, presenters, and topics jostled for inclusion. The Women's Institute movement was formed in Britain in 1915 to address the wartime food crisis. By 1925 it claimed more than a quarter of a million members. The organization saw the value of the radio for rural women and quickly took up a role in shaping the new medium. The National Federation of Women's Institutes (NFWI) chairwoman Gertrude Denman was on the BBC

Women's Committee in the early 1920s and gave the first talk of the 1930s *The Countrywoman's Day* series. Lady Denman's role as NFWI chair gave her authority, legitimating her expertise. Many other less well-known women who broadcast in the interwar era rested their authority instead on experience and common sense, something that listeners appreciated. Mrs. Webb was thus able to develop a career as a professional woman with the help of the new medium of radio.

Household Talks also provided many opportunities for the predominantly male experts of the world of science, medicine, nutrition, infant health, and welfare to come into the rural housewife's kitchen and attempt to educate her. The idea of applying scientific management to housework had spread to Europe in the 1920s following the publication of Christine Frederick's *Scientific Management in the Home*.[16] Radio programs in Germany, for instance, promoted the view that "there was no part of women's experience that could not benefit from rationalization" whether it be domestic technology or housework.[17] Working-class women's experience of domesticity was, however, framed by material circumstances and poverty, particularly when, in 1931, unemployment reached three million, an estimated 20 percent of the working population.[18] Women "struggled to balance their budgets and feed their families in the face of overwhelming odds,"[19] with varied housing conditions and patchy or limited access to water, electricity, and reliable cooking facilities. V. H. Mottram was a frequent BBC broadcaster, a dietician from the King's College of Household and Social Science, University of London, who placed much emphasis on vitamins and minerals in the diet.[20] He was also vociferous in his criticism of nonscientific broadcasters—and what he termed the "amateurism of the advice" of others, who were usually women. Nevertheless, the BBC *Talks* director noted, cookery talks seemed to "come better from a woman"[21] perhaps because their advice was generally more attuned to the limited budgets of their listeners. Mottram unleashed one of his many written diatribes when a trained nurse recommended that cheap porridge could be made with water rather than milk; he described porridge as a "fool of a food," which apparently "any but the Scots" would have discovered.[22]

Mottram continued to broadcast for many years despite his falling out with the minister of food over winter salads during World War II. Media historian Michael Bailey has suggested that all such broadcasts were a form of governance, "one of many domestic technologies whose intended use was, in part, to regulate and direct women's behavior and conduct for patriarchal reasons of state."[23] Bailey, however, ignores the contradictory and complex nature of the *Talks* programs, which, despite Mottram's efforts, did not provide consistent advice and were listened to by an audience who used their own critical facilities and experience to engage with "experts." The wireless operated as a space in which the competing discourses of science, scientific management, and rural expertise competed for legitimacy. On this democratic medium, where listeners expressed their views in numerous ways, the rural expertise of women like Mrs. Webb won out. One woman wrote to the *Radio Times* that "as a housewife I would like to say that I find your domestic talks helpful and entertaining."[24] If the talks ceased to amuse or appeal she could, of course, have switched her set off.

Mrs. Arthur Webb, christened Mabel Elizabeth Edwards in 1874, was the daughter of a farmer in Knighton, a small market town in the Welsh county of Radnorshire, very close to the English border. After her marriage in 1899 to Arthur Webb, secretary to the Co-Operative Building Society, she lived most of her life on the outskirts of London, more than 137 miles away from her birthplace. Arthur was five years her senior and born in Battersea, London, into an established middle-class family that, if not left wing, had some radical and "social conscious" leanings. The 1901 census suggests the couple were living at this point with his mother, sister, and their first child, Gwynnett, in Battersea. Three more children, Rhonna, Seymour, and Harold, were born within the next five years to complete the family, which settled in Wimbledon, a relatively affluent area on the border between Surrey and London. Mabel, alongside other well-educated middle-class women, found the time to support the Wimbledon branch of the London Society for Women's Suffrage.

The suffragists intervened when, in 1907, the Wimbledon Conservative Parliamentary Party selected Henry Chaplin, who was opposed to women's suffrage, as their candidate to become its next member of Parliament. Bertrand Russell, a philosopher and mathematician of aristocratic linage, was persuaded by his wife, an American Quaker, to stand as a pro–women's suffrage candidate for the Liberals. At the first crowded and rowdy meeting, which opened Russell's campaign, rats were released into the hall to frighten the women in the audience of approximately two thousand. Russell's autobiography refers to Mrs. Arthur Webb as one of those who spoke in support of his candidacy.[25] Following a lively campaign, with eggs thrown and disturbances at many meetings, Russell polled just 3,298 votes; Henry Chaplin, with 10,262 votes, achieved a record majority of 6,964.

By 1911 Mr. and Mrs. Webb and their children were living at Kootway, Durham Road, Wimbledon, which remained their family home for a number of years. They appear to have been comfortably off but not wealthy; they had only one live-in "lady-help," as Blanche Emma Watling from Essex was described in the census. This was a very modest level of domestic help for a mother of four young children in Edwardian Britain. Even if another woman perhaps came in to assist with the weekly washing, it would have meant that Mabel's life included a significant amount of housework. Thus, it is not surprising that in the 1911 census she is not listed as engaged in paid work. World War I changed this, enabling Mrs. Arthur Webb, as Mabel was publically known, to mold together her experiences as a farmer's daughter and her married life as a suburban housewife to become a professional rural woman, an expert on domestic production, preparation, and preservation, providing advice to those with gardens, allotments, smallholdings, and farms through many publications and broadcasts during the interwar years. Her public life began when, as part of her war service, she managed a farm and lectured on food production.[26]

The war made food, domesticity, and housewifery more culturally significant in Britain than it had been. The declaration of war in 1914 was greeted by panic buying and price rises. Some shops closed early, while others had to deal with disenchanted customers. In the first week of the war—and for the next four years—the cost of food, as Jay Winter has argued, "was a controversial subject. . . . Many voices were raised repeatedly about increased food prices and the related phenomenon of 'profiteering.'"[27] As the war progressed, the food crisis led to price rises, greater food shortages, and lengthy food queues.[28] The government attempted to cajole and persuade the population to consume less bread, sugar, and meat and to grow and eat more vegetables.[29] The effort met with some limited success.[30] The situation, however, was exacerbated when German submarine blockades increased in 1917 and then in 1918, when food rationing was introduced.[31] During this food crisis, rural women's potential contribution to food production was increasingly perceived as important. In 1916 the Women's War Agricultural Committee set up branches in each county to encourage the mobilization of women onto the land. The effectiveness of the local committees varied. Thomas George has charted the attempts to encourage more women to work on the land in Wales and noted some

Women's Institute activities during World War I, making jam and wicker baskets to help with the war effort. Courtesy of Mary Evans Picture Library / Onslow Auctions Limited.

reticence among women, especially if wartime provided other employment opportunities.[32] Nick Mansfield has also identified some married rural women in Shropshire who were unwilling to increase their labor,[33] and a similar pattern emerged in Worcestershire.[34] Arrangements were made to supply women with appropriate clothing, and "women who had worked on the land for not less than 30 days, or 240 hours, were entitled to wear an armlet of military appearance (khaki, with a red crown)."[35] A number of organizations, including the Women's Legion and the Women's Farm and Garden Union, promoted the training of women to work on the land.[36]

Women who worked on farms were considered important, but they were not the only ones who made a contribution to the nation's food supply. The *Coventry Evening Standard* remarked: "Probably there is no woman who leads a more cramped, unattractive life than the countrywoman. In most villages there are no amusements, no interests, nothing outside the little home and the garden. But the countrywoman has become a very valuable asset to the nation. But for her, in many villages throughout the country, there would be no one to care for the gardens and allotments."[37] To support and encourage these rural housewives, the Women's Institute (WI) movement was formed in Britain in 1915, backed by the Board of Agriculture with the aim of encouraging domestic food productions and preservation.[38] The wartime WI emphasized that food production and food saving was a patriotic duty. Members were encouraged to ask themselves "What is my home, my garden, my farm doing for my Country?"[39] and to undertake rabbit and goat keeping, gardening, cheese making, herb collecting, poultry keeping, jam and pickle making, fruit drying, and food economy.[40]

The Board of Agriculture funded lecturers speaking at WI meetings, and training centers also supported the expansion of colleges to teach women agriculture or horticulture;[41] all provided opportunities for women like Mabel Webb. She drew on her background as a farmer's daughter and her experience as a suburban housewife when giving talks; she was on the Board of Agriculture's panel of lecturers in Hampshire, Surrey, and Sussex during the conflict; and her expert reputation grew particularly in relation to the preservation of fruit and vegetables, leading to her appointment as a county judge at agricultural and horticultural shows.

If a growing professional persona was Mabel's personal wartime legacy, then other legacies of the conflict facilitated her building a career. World War I reinforced the concerns about motherhood and domesticity that Anna Davin has argued emerged at the end of the nineteenth century.[42] The poor health of recruits for the Boer War and World War I accelerated the criticism of working-class women for their "lack of ability as household managers, and most especially for their lack of cooking skills, in selection of food."[43] Hence, as Judy Giles points out, there was an increasing concern over "national efficiency" and calls for motherhood and housework to be considered "as important to the survival of the nation as advances in war technology or industrial production."[44] Women's organizations, such as the Women's Cooperative Guild and the Women's Institute, mother and baby clinics, and the media were all identified as potential spaces in which domestic skills could be both taught and learnt, and women like Mabel Webb were needed to do the teaching in wartime and the years that followed. Webb involved herself in domestic and agricultural education,

Women canning apples on the home front during World War I. In 1917, before rationing was instated, a food economy campaign urged the public to save food and to cook. Courtesy of Illustrated London News Ltd / Mary Evans Picture Library.

lecturing to WIs and horticultural societies, including the Vacant Land Cultivation Society, and by 1930 had begun to publish cookery books.

Webb's broadcasting career began on January 11, 1932, when she gave a talk on marmalade-making for the BBC's weekly early afternoon program, *Hints from Other Cooks*; she returned in March to broadcast on batter before speedily moving on to become a regular contributor to *Household Talks*, covering subjects including honey, fruit bottling, jams, crystalizing fruits, and supper dishes. Food preservation and production—the traditional focus of rural housewives and Women's Institutes—were her key areas of expertise. In September that year, the *Radio Times* announced that "Mrs. Webb will give practical advice on how to catch and then preserve the bilberry, the whortleberry, the sloe, the wild raspberry, the wilder strawberry, the elusive blackberry, and the shy cranberry. This, the first of a series of four morning talks on spare-time cookery, should convert England into a nation of bottlers and jelly-makers. Simple recipes will be given, all within reach of the shortest purse."[45] Webb seems to have naturally adopted the informal, intimate mode of address that radio required and the frugality that the housewives' budgets needed. In 1933 she broadcast *Economical Cooking*, a series of twenty-four talks. At a time of economic crisis, her approach to food echoed measures used during World War I: using local or homegrown vegetables and fruit, avoiding waste by food preservation, and preserving jams, chutneys, jellies, pickles, and marmalades. Her advice was aimed at rural housewives, for whom it was particularly useful, but it also appealed to many others who identified with rural ways of life.

There are hints in the archives that the BBC did not always receive Webb's broadcast scripts quite as far in advance as they would have liked and that the titles of her talks sometimes arrived so late that they could not be included in the program listings in *Radio Times*. In 1933 the BBC suggested that listeners needed a wider variety of speakers. Her tone with the BBC in the following months can only be described as groveling, even when she asked to use some of the images and recipes to create a book, which was published as *Mrs Webb's Economical Cookery* in 1934. The cookbook was down-to-earth, grounded in British rural food. She included recipes for pot roasts, pig's head and tail soup, and how to cover half a pig's head with pastry to make a roast. The recipes were suited to the lifestyle of rural families who kept a pig in their backyard or garden and for whom homegrown fruit and vegetables made a significant contribution to their diets.[46] Webb maintained the familiar conversational style that made her a popular radio broadcaster and ensured that the accompanying leaflets that the BBC produced sold well. The BBC sold 370,506 *Household Talks* pamphlets between 1933 and 1935, by post and through Women's Institutes, including her 34-page booklet *The New Herring Book* and *The Wise Penny 2—Hints on Economical Marketing and Recipes*, written with Emelie Waller in 1934. Her guidance on *Jam, Jam Making and Fruit Cheeses* began informally: "Do you make strawberry preserve—leaving the fruit whole? I always think it looks nice and tastes good. Many clever cooks think it better to do it this way. It certainly takes longer and requires more care when syrup has to be made first before the fruit is added. As I know, many have not always got the time to spare."[47] By this time Webb's publishing career was blossoming. She had published articles in the *Listener* and *Radio Times*; she also became the cookery editor of *Farmers Weekly*, a magazine established in 1934 for the British farming industry, that described itself as a "newspaper of the soil."[48] Throughout the 1930s she bombarded the BBC with suggestions for programs and gave BBC personnel the products of her garden and kitchen, such as pears and chrysanthemums, apple mould, parsley, and roses, in the hope that these would smooth the way to further radio appearances. She was rewarded with a large number of broadcasting opportunities covering subjects such as "dealing with bumper apple crops," pot roasts, tasty ways with bacon, and potato surprises.

Two elements of her broadcasts explain Webb's popularity: her style of addressing the public, presenting herself—in educational terms—as a "guide at the side" rather than a "sage on the stage";[49] and her easy identification with an idealized rural home, which had become more culturally significant during World War I, when the ideal of England, the nation that men were fighting to protect, was perceived to be rural. It was an England idealized by the Edwardian artist Helen Allingham in her paintings depicting rural cottages with contented wives and children waiting at the door. The images appeared on the postcards that soldiers sent home from the Western Front during the war. A recruitment poster portrayed a British soldier exclaiming to the public: "Isn't this worth fighting for" as he points toward rolling countryside, a dairy herd, and, most important, a thatched cottage.[50] The urban working-class communities that many soldiers came from bore little resemblance to these idealized images. Yet the first industrialized war reinforced and exaggerated the identification of Englishness with rural life, which was perceived in utopian terms in contrast to the harsh consequences of urban industrialization.[51] Men and women longed for ordinary domestic life at the end of the war as "the ultimate retreat from the disciplines of military society."[52] As both the director of the Ministry of Pensions and the London *Times* acknowledged, home was where soldiers wanted to return.[53]

By the 1930s, it was clear that for many, everyday life did not match the ideal for which men had risked their lives.[54] According to R. J. Moore-Colyer, writing in 1999, "at innumerable industrial towns between Bradford and the Black Country, the Victorians had made their money and left their muck, so that by the early 1930s the fag-end of Victorian progress was manifest in bleak, dank streets overhung by unsavoury smog and populated by despairing communities of unemployed."[55] For many in urban areas, neither workplace nor home fulfilled their hopes and dreams. They sought to escape permanently into new suburban developments built on farmland around larger towns—or temporarily, by visiting the countryside. Groups such as the Co-operative Holidays Association, Cyclists Touring Club, Camping Club of Great Britain, Youth Hostels Association, and organized walkers, including the Ramblers Association, provided opportunities to visit the countryside.[56] Thus, rural Britain became, in the British imagination, a site of leisure. Moore-Colyer has argued that even in "the 1920s the British rural idyll, or more strictly, the *imagined* rural idyll, was coming under threat from a combined onslaught of people, roads, rural electrification, unplanned urban spread and myriad other stresses."[57] The rural in many ways belonged in the past, when Webb had spent her childhood on a Welsh farm.

Webb's various broadcasts on *Household Talks* connected to the rural life of the imagination of many urban dwellers as they sat down for their morning break listening to the radio. How many of the housewives or servants who consumed her slightly nostalgic broadcasts and cookbooks actually tried to make the recipes themselves remains a matter for conjecture. Food operates as a signifier of a way of life, values, or moments in time in national, personal, or family histories and memories.[58] Thus, when Webb visited the London Master Baker's Exhibition in the 1930s and reported on the event to her listeners, she spoke of her childhood, her father's farm, and a rural domestic idyll that her listeners understood and enjoyed: "I reached the seed cake section. The spicy caraway scent was there reminding me of great chunks of cool moist cake eaten years ago in the dairy. If you forget about the slimming, try farm seed cake and fresh cream. But to really enjoy it you must have risen early and had an early breakfast."[59] Her listeners would have warmed to this presentation of herself as a hard-working farmer's daughter and appreciated the informal and intimate style she used when she also positioned herself on the boundary between "ordinary housewife" and "extraordinary" rural expert to become a friend at her listeners' firesides. On other occasions, Webb spoke on behalf of ordinary housewives, mediating "between expert speaker and the lay public," an approach adopted to make experts more comprehensible to listeners.[60] In 1934 she interviewed representatives of the new agricultural marketing boards for milk and bacon. When she questioned a representative of the Milk Marketing Board, she asked whether the farmers and distributors had given due consideration to housewives and whether surplus milk should not go to school children rather than factories for the production of milk-related products.[61] Three year later, John Boyd Orr's study, *Food, Health and Income*, emphasized the link between children's growth and nutrition and recommended an increase in young children's milk consumption,[62] just as Webb had suggested to the Milk Marketing Board.

In 1934 the BBC took the radical step of allowing "ordinary housewives" to broadcast themselves, giving advice to other women in a series called *How I Keep House,* in which they explained how they managed their budgets. The series included the wife of a Norfolk farm laborer and a smallholders wife, whose experience the *Radio Times* summarized: "How I keep house by smallholder's wife—her and husband Londoners—She will describe how they lost their money on poultry and had to mortgage their property in order to carry on. A hard life. No fixed wage coming in. All the work entailed in summer visitors to eke things out. Stock to feed, no modern conveniences, it says much for the country that, in spite of all, she has no wish to live in town again."[63] These women presented a less idealized version of rural domesticity than Webb's broadcasts did, but the advice was just as frugal. Mrs. Betts, the wife of an agricultural laborer, explained how she made a hot dinner every day with a pudding to save on the purchase of bread and as her "husband doesn't feel he's had a dinner without a pudding." Her strategies to ensure her family's survival echoed many of Webb's suggestions, including picking blackberries and crab apples to make into jam for the winter, sewing her children's clothes, limiting use of her oil stove, her husband's bee-keeping, and tending their allotment. She further described how her Sunday joint lasted four days. She made "suet puddings flavoured with the gravy set aside from the meat," a thrifty practice that, along with reheating meat, Mottram was known to criticize. Mrs. Betts went on to explain other tasks: "We have to carry all our drinking water a distance of 60 yards but my husband does that for me and we catch our washing water in a tank outside the back door. All dirty water has to be carried out on to the garden a distance of 20 yards, a bother on a wet washing day."[64] Allowing ordinary working-class women to describe how they coped with their hand-to-mouth existence was "greatly prized" in working-class women's culture[65] and blurred the distinction between the professional rural woman and the housewife. The popularity of *How I Keep House* may have influenced the planning of *Farmhouse Cookery.* In making the series, Mrs. Webb explained to her listeners, she had encountered "the real spirit of adventure in cookery. Fearlessness based on skill and knowledge, a flair for producing the right effects, the perfect flavours, judgment and tradition asking no help from modern appliances. . . . Once a week, bread and such bread—plain, spiced, currant, wholemeal, Yorkshire teacakes, pies, cheese cakes, tarts, and great big useful fruit cakes, by magic from the work of that pair of capable and utterly efficient hands."[66] Webb's role was to transmit the "natural" and even "innate" skills and expertise of farmers' wives to listeners while also serving as the ordinary housewife's representative:

> I looked at the fireplace, watched the flames travelling under the oven, "How do you manage to keep the heat going, you usually burn coal of course?" "Oh no," the answer came swiftly. "Oh dear no, I never trust coal, or anything else than wood for my baking. I understand wood better, and know exactly what heat it will give."
>
> "Do you have failures?"
>
> "Failures? Of course not, I know exactly what I want to make and make it."[67]

This style of rural food preparation was far removed from the scientific approaches to housewifery that male experts such as Mottram advocated. For instance, Webb asked:

> "Well how do you manage to arrive at such delicious pies as these? Do you weigh the ingredients?"
>
> "Never. I could not spare the time. I just know how much the flour butter, lard, milk, water and eggs will make."[68]

This interchange, however, does not come from a recorded interview. Rather, Webb visited the various farmhouses and later wrote the scripts and then recorded the talks—for which she received a fee of seven pounds for the series.

At a point when, for the first time, "millions [were] all listening to the same program," creating the "sensation of being part of the nation in a way that was rarely experienced in the past,"[69] the *Farmhouse Cookery* broadcasts and the accompanying book constructed a sense of both regional and national identity. Place was initially conveyed through descriptions of the landscape, for example: "In going from farm to farm in Kent and Sussex I have been thrilled with the old-world buildings and the unspoiled beauty of the counties. It is not only the orchards of Kent and the ploughed lands of Sussex, but the delightful gardens and fine woodlands, which give the traveller the idea that London is hundreds of miles away."[70]

Then there were descriptions of architecture, decor, and artifacts in the farm kitchens, the different elements combined to create a regional geography for the imagination. The food and recipes were strongly regional in nature, linked to the landscape in which they were cooked: Quantock Pigeon Pudding, St. David's Stew, Cheddar Pork Pie, Yeovil Veal Loaf, Old Sussex Meat Loaf, Montgomery Fruit Scones, Scotch Broth, and Keswick Lemon Pudding. Linking the food to the landscape in Devonshire, Webb explained that "as I saw the cows knee deep in thick meadow grass, I was not surprised that this was the land of cream and butter." The description preceded a recipe for Devonshire Chudlieghs, which are eaten with jam and cream.[71] The nostalgic idea that the best of rural food and housekeeping belonged perhaps to a world that was under threat and slipping away is also evident in Mable Webb's introduction to a recipe for Swanley Harvest Custard. She explains that it was the farmer's wife's most popular recipe; "her mother had made a special harvest dish which was carried cold to the family working in the fields, or made fresh and hot for supper after the day's labours."[72]

The series and the cookbook both sang the praises of regional particularities and specialities in food while joining them with a sense that the heart of a nation was the rural home.[73] Shropshire was apparently a county possessing skilled cooks,[74] while Leicestershire, "the proud home of the wonderful Melton Mowbray pork pie," also provided an unpretentious recipe for a royal sandwich, which consisted of two slices of homemade bread with ham and marmalade between them. The down-to-earth, folksy nature of *Farmhouse Cookery* cemented Mrs. Arthur Webb's expertise in rural food production, preparation, and preservation. Despite the support of a secretary, Mr. Lukor, whom she now employed, the punctuality of submitting her scripts continued to be erratic. Nevertheless, the BBC used her to review other broadcasters' scripts and give advice on the recipes. In 1936, when the BBC held a conference on the *Household Talks* program, she was an obvious choice to lead the morning discussions, because the many members of Women's Institutes who attended ensured rural women were strongly represented.

In the year following the publication of *Farmhouse Cookery*, Mabel Webb broadcast from the Ideal Home Exhibition at Olympia, the Imperial Fruit Show, the Ice Cream Exhibition at the Crystal Palace, the British Industries Fair, and the Early Vegetable Show at the Agricultural Hall in London, widening her listeners' engagement with both food production and preparation. Her broadcasts certainly made a contribution to the claim made in 1939 that "the whole attitude to housekeeping and motherhood is undergoing modification in the direction of increased knowledge, control and dignity, here again broadcasting is not the sole agency at work but it is one of the most effective."[75] This might perhaps overstate the influence of broadcasting, but by then there were nine million license holders and "regular audiences significantly larger than" the number of license holders[76] whom Mabel Webb was keen to address. In August she wrote to the BBC Talks director, Miss Quiggley, hoping for more commissions:

> I am feeling quite sad; here is the last talk of the series, and I have so thoroughly enjoyed giving the broadcasts. The work still remains but will be got through in due course (re letters and sending out receipts etc)
>
> If the apple crop should come tumbling down in September and early October it might be worth while to deal with methods that will account for quite an amount of preserving, but that must be left to you.
>
> With greetings and all good wishes
> And many thanks
> Believe me sincerely[77]

By September, however, the challenges of the apple crop had given way to those of war. As World War II grew imminent, Webb again sought to take advantage of the opportunities it offered and was able to broadcast her first *Making the Most of the Wartime Larder* programs on September 8 and 12. She also published a wartime cookbook before the end of 1939; a second edition followed in 1940.

Webb continued to broadcast during the war. Although she briefly moved to High Wycombe during the bombing, she returned to her house in Battersea in 1941 and contributed to a series called *The Kitchen in Wartime*. Initially, she was even asked occasionally to broadcast in the better paid evening slots, once the BBC had acknowledged that more housewives were working in wartime. Her talks, such as "Herrings for Breakfast," brought requests for recipe leaflets; thus, she went on answering letters from listeners and commenting on other broadcasters' scripts. By the end of 1941, however, things were changing. The BBC was subject to a high level of government "guidance" during wartime. The Ministry of Food often vetted scripts and encouraged the BBC to make their domestic advice more appealing by using music hall celebrities and comediennes to present them. The war also provided opportunities for a range of younger broadcasters, with a greater emphasis on modernity and looking forward to building a new urban and suburban postwar world.

The indefatigable Mrs. Webb's career spanned two wars, and she continued broadcasting into the 1950s. She had propelled her career as a rural woman with determination and hard work, and by pestering BBC executives and offering them examples of her rural produce. Her popularity also relied on her appeal to both rural and suburban audiences. She was not only an expert but a conduit for the expertise of others as she took advantage of the possibilities that the cultural legacy of World War I offered her.

8

African American Home Demonstration Agents in the Field and Rural Reform in Arkansas, 1914–1965

CHERISSE JONES-BRANCH

Between World War I and the civil rights movement, Southern home demonstration agents, employed by the Agricultural Extension Service and informed by progressive ideas, targeted farm women and promoted domestic reform in rural spaces. On the surface at least, it appeared that home demonstration agents, regardless of race, were charged with improving conditions in rural environments. African American home demonstration agents in Arkansas, like black agents around the South, however, assumed the additional burden of negotiating deeply entrenched Southern racial codes. Although they helped rural blacks circumvent the harsh reality of their lives by imparting farm and home economics techniques as tools for self-sufficiency, black home demonstration agents further crafted strategies to alleviate African Americans' endemic poverty, food insecurity, and unsanitary and unsafe living conditions. Working cooperatively with rural black Arkansans, agents employed by the Arkansas Agricultural Extension Service ably navigated the constraints of racism, segregation, and severe privation to reduce economic marginalization and to increase African Americans' access to health care and food security.

Home demonstration work in Arkansas was supported by the 1914 Smith-Lever Act, which created the U.S. Department of Agriculture Cooperative Extension Service.[1] Informed by Progressive era concern for Southern rural reform and development, the act supported, using federal, state, and local funds, the establishment of an agricultural extension service at state land-grant colleges and universities. The extension service was established in Arkansas shortly after the passage of the Smith-Lever Act and was headquartered at the University of Arkansas in Fayetteville (Washington County) for whites and at Arkansas Agricultural, Mechanical and Normal College in Pine Bluff (Jefferson County) for African Americans.[2]

Extension service policies reflected and reinforced the traditional gender divisions of the early twentieth century. Because of the erroneous assumption that only men engaged in farm production, the extension service hired and supervised farm agents, who were almost exclusively male, to advise farmers in crop production measures.[3] The extension service also employed female home demonstration agents, but the assumption was that they would only work with farm women on home economics projects.[4] Arkansas's first agent, Harvey Cincinnatus Ray, more commonly known as H.C., was not hired until 1915. A graduate of Lincoln Institute, a predominantly black school in Jefferson City, H. C. had been a faculty member at Langston University, also a black institution, in Oklahoma.[5] His wife, Mary Lee McCrary Ray, an 1897 graduate of Tuskegee Institute whom H.C. had met at Langston, had been head of the home economics department at the university.[6] In 1916, she was appointed Arkansas's first black home demonstration agent. In July 1918, she was promoted to district agent, with responsibility for all black home demonstration agents statewide. She held this position until her untimely death from chronic interstitial nephritis and arteriosclerosis in 1934.[7]

The extension service reinforced Jim Crow laws mandating racial segregation.[8] Black demonstration agents reported to white supervisors and dared not challenge local racial strictures. In Arkansas, their education marked them as potential agitators. White landowners were often initially suspicious of black agents because they feared they might disturb the carefully crafted social and economic control they held over their laborers.[9] Particularly in the Arkansas Delta, where African Americans constituted a majority, black and white agents were loath to sacrifice their programs by overtly challenging white supremacy.[10] H.C. and Mary Ray, for example, traveled throughout Arkansas demonstrating farming and canning, and giving lessons in euthenics, a new discipline that emerged at the turn of the twentieth century to address environmental effects on race.[11] But as they did so, they carefully cultivated the goodwill of planters and county officials. This allowed them to maintain a force of farm and home demonstration agents in areas with a large concentration of blacks who desperately needed and desired their services.[12]

African American home demonstration agents had been working in Arkansas since 1913.[13] America's entry into World War I soon increased the need for their services around the state. Agents were concerned about cash-strapped homemakers whose families faced food insecurity and the resulting dietary deficiencies. Cotton remained a primary crop for many farm families, who as a result were often unable to grow vegetables to supplement their meager diets during the winter months. Throughout the war years, farm and home demonstration agents urged black Arkansans to increase food security by planting gardens and canning produce.[14] Mary Ray, for example, helped farm families turn canning tomatoes into a community project. Using facilities that were typically owned and/or operated by African Americans as well as other available resources, rural blacks produced five hundred quarts of tomatoes daily in a cannery built out of zinc tubs, with a foundation, top, and the joints of two stovepipes.[15]

Although local funds were limited during and after World War I, white landowners increasingly depended upon agents' services to stem the tide of black outmigration. By the summer of 1918, Southern states had procured additional home demonstration agents to visit black homes and farms.[16] Their work was encouraged as a means to help overcome what one agent called "the poverty, dissatisfaction, and unrest among Negro farmers" who resisted the oppressive conditions they endured in the rural South.[17] Like their enslaved ancestors, rural laborers questioned and contested the racial, economic, and political mechanisms that held them captive, often engaging in overt and covert acts of resistance. African Americans used their limited power to gain what they could from their employers, including lying, cheating, and dissimulating when necessary.[18] When all else failed, African Americans resisted with their feet and simply left in search of the social, political, and economic access denied them in the South. Beginning in World War I, six million blacks quit the South as part of a migration that did not end until the 1970s.[19] Rather than perceiving black agents as a threat to the plantation system, white landowners, panicked by the loss of their cheap agricultural labor force, increasingly realized that it was in their best economic interests to embrace the Arkansas Agricultural Extension Service's reform efforts.[20]

Home and farm agents further cooperated with social workers and black community leaders to consider ways to make farm life as productive as possible and to secure as many concessions as possible for agricultural laborers to encourage them to remain in the South. One of the groups the agents worked with, for example, consisted of the Jeanes supervising teachers, who were funded by an endowment left in 1907 by Pennsylvania Quaker Anna T. Jeanes to support African American education in cooperation with state and county school officials.[21] In 1925, for instance, the Rays and a delegation of farmers attended the annual farmers' conference at Tuskegee Institute, where black leaders discussed educational, health, and, of course, agricultural concerns among rural farm folk.[22]

By 1921, 237 African American extension service agents were employed in fifteen Southern states. In Arkansas, black county home demonstration agents were located in Conway, Desha, Faulkner, Lee, Mississippi, Phillips, St. Francis, and Sebastian counties.[23] Though their numbers were small, their presence was critical because they impressed the importance of food preservation on rural women, particularly as rural economies began sinking into a depression. In 1924, black women and girls who were members of home demonstration clubs showcased clothing, poultry, and fruit and vegetable preserves, reflecting an idealized gender division of labor based on the assumption that farm women focused solely on home-based projects, although that was seldom the reality. Even so, black women and girls often won prizes at the Pulaski County state fair for their work, which they proudly displayed as examples of the educational benefits they received from agents and home demonstration club membership.[24]

Although food insecurity certainly affected poor, rural black families in the years leading up to the Depression, many were first devastated by the Mississippi flood of 1927. Floodwaters from the Mississippi River covered four million acres in eastern Arkansas and resulted in the destruction of millions of dollars of property and the loss of 127 lives. The flood further revealed the extreme poverty of those, most of whom were African American, who lived in the Mississippi River Delta region.[25] Working with the American Red Cross, black home demonstration agents' and club members' skills helped provide food and housing for destitute but proud African Americans as they struggled to survive the storm. Relief worker Lula Toler, an agent from Pine Bluff, Arkansas, recalled the suffering she encountered: "My first day's work with the sufferers were ones long to be remembered. The scene was most pitiful and shall ever haunt me while life lasts. The screams of the people and the lowing of cattle could be heard in the distance."[26] Toler, who resigned her post in 1927, had been the Jefferson County home demonstration agent since 1925, the year during which she also established its local council. Among their many accomplishments under her leadership, club members who had been trained as "health leaders" helped their communities access important health-care opportunities.[27] Toler had previously been a home demonstration agent in Meridian, Mississippi, and had also been a high school science teacher in Birmingham, Alabama.[28] After resigning from the extension service, she became a noted ordained minister and the founder of the first convalescent home for African Americans in Pine Bluff.[29]

St. Francis County agent Lugenia Bell Christmas assisted those who had waded through floodwater for miles in search of food and shelter for their families.[30] Christmas was not alone in her efforts, however; local home demonstration club members distributed food to those who were in an emergency hospital that was led for three weeks by Forrest City club leader Henrietta E. Ankrum.[31] Agents never missed an opportunity to teach flood victims about domestic hygiene. The American Red Cross placed Lee County agent Annie L. Smith in charge of securing sanitary "sweet milk" for black infants. Providing children with sanitary milk was important because rural Arkansans, regardless of race, often lived in impoverished environments and ingested unsanitary and unsafe milk, which led to increased child mortality rates. At times, this was attributed to milk that had been obtained from unhealthy cows. At others, the cause was unclean utensils, buckets, vessels, and separators that had been used to handle the milk. Extension service agents warned against the dangers of such milk-borne diseases as typhoid fever, tuberculosis, scarlet fever, diphtheria, and septic sore throat. They further stressed the importance of serving children fresh, cold milk to avoid outbreaks of dysentery and diarrhea.[32]

Smith also attempted to combat food insecurity among African American flood victims by organizing home demonstration club members into four "canning club units" that were then responsible for holding canning demonstrations in their assigned section of the camp. Afterward, black flood victims were sent to pick and can blackberries. Using canneries, cans, and sugar provided by the American Red Cross, they produced several hundred quarts of canned berries.[33] Such efforts not only improved food security but also combated pellagra outbreaks, which was particularly important because people in rural areas, regardless of race, endured diets tied to a cash-crop economic system. As tenant farmers or sharecroppers, they were unable to grow much of their own food and often suffered from pellagra because of insufficient protein and niacin in their diet. This initiative had long been an important part of home demonstration agents' rural activism and became even more so during the 1927 flood relief efforts.[34]

It is quite likely that the overwhelming responsibility and strain of helping flood victims and bearing witness to the devastating poverty and the extreme racism they endured proved too much for some black home demonstration agents. Close to the end of the 1927 flood, Annie L. Smith resigned, just as Lula Toler had earlier. White home demonstration agent for Lee County, Flora Ferrill, subsequently assumed responsibility for organizing canning clubs for African Americans living in what was euphemistically called "the bottoms," the separate and certainly unequal space designated for black flood victims.[35] Black home demonstration agents appreciated their white colleagues' assistance. As district home demonstration agent and director of Negro work during the 1927 flood, Mary L. Ray noted this important professional, if not quite equal, exercise in interracial cooperation while highlighting black women's accomplishments under difficult circumstances. In a 1927 report, she credited black women all over Arkansas with improving food availability and security by producing more than 130,000 quarts of canned fruits, vegetables, and meats. But even as she praised white agents' and officials' efforts, she reified their racist and her own classist beliefs about rural blacks' backwardness and inferiority by asserting that they "have always shown their interest and faith in the future of club work and its effectiveness to advance a backward people."[36]

Nonetheless, home demonstration agents were critical allies during the flood crisis because they helped victims find lost loved ones, organized soup kitchens, and provided lessons in hygienic domestic arts. In tandem with male farm agents, they also conducted agricultural demonstrations and provided landless and displaced victims in Arkansas relief camps with much-needed psychological support as they rebuilt their homes and lives.[37]

Black home demonstration agents' domestic reform efforts reflected Booker T. Washington's philosophy of black self-sufficiency and determination. In 1928, the Arkansas Agricultural Extension Service established the Movable Demonstration School for Negro Farm Folks.[38] Movable schools, developed in the early twentieth century by George Washington Carver and Booker T. Washington, were agricultural schools on wheels that moved from one community to another, similar to the practice in the Netherlands that Margreet van der Burg describes in chapter 3. Black extension service agents traveled throughout Arkansas in a one-and-a-half-ton Ford truck and performed demonstrations in approximately twenty-eight to thirty counties in the first year of the school's existence.[39] Access to movable schools allowed the most isolated rural black farm families to learn about the latest agricultural, home management, and public health innovations.[40]

As the demand for home demonstration agents grew across Arkansas, so too did the need for a statewide organization to unite and coordinate rural women's activities. Because farming families often lived far from their neighbors, the clubs provided opportunities for women to socialize.[41] Female-led home demonstration clubs had a positive impact on rural communities. They were safe, women-centered spaces to inquire about family, health, and child care concerns. These clubs also combined educational programs with the extension service's agenda and cultivated relationships with current local leaders and created new ones. They provided additional important spaces for agents to educate rural women, who were then expected to share their newfound knowledge with others.[42] Although women may also have engaged in political discussions during club meetings, such an undertaking would have been extremely dangerous in rural Arkansas, where landowners, who needed a compliant workforce, possessed considerable power and influence over their black agricultural laborers.[43]

In 1929 white home demonstration agents organized the Arkansas Council of Home Demonstration Clubs.[44] Jim Crow restrictions prohibited black women's membership. But black women worked together to create a state organization of their own. In 1936, the State Council of Home Demonstration Clubs was founded at Arkansas Agricultural, Mechanical, and Normal College. Like the Arkansas Council, the State Council created a network of county clubs in rural black communities around the state to support domestic reform activities.[45] The State Council was founded in the mid-1930s, but black home demonstration clubs in Arkansas had been around since the mid-1920s. Their numbers increased during the Depression years, however, because many black families were so impoverished that it was imperative for agents to teach them new farming and home techniques that allowed them to improve agricultural production, generate much-needed cash income, and become self-sufficient.[46]

Like white women's clubs, much of black women's home demonstration club programming in the 1930s involved canning. Starting in 1932, approximately twelve hundred canneries or canning kitchens and sewing projects were created. These groups, subsidized with funds by the Civil Works Administration, put many of the women back to work in areas such as North Little Rock (Pulaski County), where the first canning program was established.[47] Skillfully harnessing women's control over household finances and food production, extension service agents often appointed club members as supervisors of canning facilities.[48] In at least one instance in 1935, a sixteen-year-old African American woman supervised a canning kitchen in Pine City (Monroe County). Cynthia Weems from Holly Grove, a ninth grader at the laboratory high school at Arkansas A. M. and N. College, not only had the distinction of being one of the youngest supervisors but her cannery produced 5,196 cans in 1934, more than any other kitchen, black or white, in Monroe County. Weems was further impressed by black home demonstration agents' leadership abilities. Agent Cassa Lawlah also recognized these abilities in Weems and along with the white county field supervisor and other black women in the Pine City community, so encouraged her ambitions that she too desired a career with the Arkansas Extension Service. Weems, who had always enjoyed canning and cooking and described it as a "hobby," asserted: "My aim in life is to finish school and be a county home demonstration agent like Miss C. L. [Cassa Lawlah] Hamilton and from that to [become] the state agent of Arkansas."[49]

Many of the women who worked in cannery kitchens had previous experience with this enterprise because, like Weems, they had been members of 4–H or local home demonstration clubs. Through their employment, these women not only helped others who were in similar circumstances but also generated a limited, but much-needed cash income.[50] Such skills also readily lent themselves to the inculcation of the ethic of community improvement and self-sufficiency among rural blacks.

To be clear, home demonstration agents were aware of the environments in which they operated. They were also mindful of African Americans' need to demonstrate their own agency and autonomy through their membership in home demonstration clubs. Local support for extension service programs was of critical importance. Rather than positioning themselves as representatives of the federal and state government sent

Corn-canning demonstration, Jefferson County, Arkansas, 1930. Most of the participants were female, with the exception of the man standing on the right in the second row. Some were also children, thus underscoring the importance of food production and security for everyone who resided in rural communities. Records of the Extension Service, Record Group 33, National Archives and Records Administration–Southwest Region, Fort Worth, Texas.

to work on poor rural blacks' behalf, agents instead cultivated partnerships with them and were compelled to respond to their needs and interests even as they sought to uplift them from their impoverished circumstances. This tactic held great import for home demonstration and farm agents and Arkansas black leaders throughout the 1930s and beyond who often preached what Evelyn Brooks Higginbotham termed the "politics of respectability" in African American communities. Club involvement also helped black women to develop leadership skills.[51] Home demonstration clubs in Desha County, for instance, held leaders' training meetings in January and February of every year. Such activities did not overtly challenge white authority; rather, they allowed black women to safely combat assumptions about African American inferiority and pathology.[52] Leadership training further allowed poor rural blacks to exercise confidence, dignity, and pride at a time when their psyches were under siege.[53]

Despite nationwide wartime mobilization in the 1940s, blacks still suffered disproportionately from the economic privation of the Depression years. Black home demonstration agents focused on teaching farm women how to improve their homes, families, and communities by doing more with less. In 1940, local home demonstration agents in Pine Bluff sponsored a Cotton Dress Revue as part of a weeklong community-wide Better Homes campaign, which included a showcase of home improvement activities.[54] The national Better Homes Movement, a domestic reform campaign, existed between 1922 and 1935 and was led by President Herbert Hoover, who was also president of the Better Homes campaign, *Delineator* editor and journalist Marie Meloney, and Harvard University professor James Ford.[55] In Arkansas, local white Better Homes committees were established in 1924. None existed for African Americans until 1929, when Better Homes campaigns were organized in six communities in Pulaski County and a Jeanes supervisor in Crittenden County organized a cleanup campaign and school and home improvement events.[56] This initiative was just as important to black women as it was to white women and perhaps even more so because it encompassed all of the projects that black women leaders were engaged in to address conditions in rural communities and schools and to counteract negative stereotypes of African Americans. For example, when the predominantly African American Ouachita County Teacher's Association met at a segregated high school in 1930, its members, led by Jeanes supervisor Cleo N. McDonald, created committees for Better Health Week, Better Home Week, Better English Week, and Negro Week (presumably to celebrate black history).[57]

The Better Homes Movement's goals were to improve housing quality and to encourage homemakers to purchase consumer items. In Arkansas, rural clubwomen (members of home demonstration and federated clubs) and extension service agents formed local Better Homes committees and created "ideal room settings in demonstration homes."[58] In Arkansas, this "ideal" was cost prohibitive for rural dwellers regardless of race, but despite this, African Americans and their leaders eagerly embraced Better Homes programs as part of their rural uplift initiative because the programs were tied to national concerns about black domestic life as a reflection of racial progress, or the lack thereof.[59] Rural black homemakers and leaders in St. Francis County were elated when, in 1933, they were recognized nationally with a special merit award from the New York City–based national Better Homes in America program:

> This award was made primarily for the instructive educational tours conducted by your committee and for the extensive work in home improvement and the widespread participation in contests. St. Francis County, under your [Lugenia B. Christmas, local home demonstration agent] direction, has developed a significant community campaign for home improvement and one well adapted to meet the needs of the present economic situation. I wish, therefore, to extend to you personally and to all of the members of your Committee and co-operating organizations, our sincere appreciation of all that has been done.[60]

It also appears that the Better Homes movement among African American women lasted beyond 1935, which is when the national program ended. In 1939, home demonstration agent Cassa Lawlah noted that 10,084 farm families had participated in community cleanup campaigns during "Better Homes Week," including building or repairing 1,437 toilets.[61] Better Homes clubs existed in black communities all over Arkansas. In 1941, black home demonstration and 4–H clubs celebrated Better Homes Week in Desha County. Its members conducted Better Homes tours in the Watson, Bayou Mason, Halley, and Kelso communities.[62] In 1942–1943, Pulaski County agent Elora A. Maxwell was appointed the county Better Homes Negro chairman. African American home demonstration agent supervisor Fannie Mae Boone, an Arkansas Agricultural Extension Service employee since 1929 who had been trained at the Kansas Industrial and Educational Institute and later at Tuskegee, was elected as the state Negro chairman. In this same year, the state Negro committee was awarded a special merit award from the Better Homes in America national office.[63]

Racial advancement packaged as reform also meant improving conditions in and around domestic spaces as well as among those who lived there. In 1943, the Lee County Home Demonstration Council encouraged rural homemakers to make mattresses stuffed with cotton as part of a program initiated by the USDA.[64] In keeping with the patriotic theme of the war years, black women also constructed "Victory garments" from flour sacks in addition to demonstrating pride in their homes and their country by planting shrubbery.[65] They also taught local women how to sew and make clothes using fifty-pound flour sacks. The women tried to make clothing out of similarly colored sacks. When they could not do so, they cultivated their own sense of fashion, independently of home demonstration agents' influence.[66] The agents may have wanted rural black women to improve the quality of their clothing—and hence their physical appearance—and to refrain from creating items that were, according to historian Blain Roberts, "seen as loud, ostentatious, and unsophisticated."[67] Such negative perceptions were often mapped onto African Americans, leading home demonstration agents "to spend much of their time convincing rural women that their clothing . . . betrayed the race," and undermined racial progress.[68] When the situation warranted, however, women depended on their own skills to make the best use of limited—or in this case mismatched—resources that suited their rural aesthetic in ways that often pushed back against home demonstration agents' conservative proscriptions.

The patriotism of the war years also served as an impetus to encourage black farm families to help the USDA meet national food production goals. In 1942, black home demonstration and extension agents, ministers, educators, and farmers met in Little Rock, Forrest City, and El Dorado (Union County) to discuss how farmers could use the features of the Agricultural Adjustment Act to implement the USDA's Food for Freedom program.[69] The program's plenary sessions stressed that rural black communities should demonstrate their patriotism by participating in local programs to increase food production and conservation. An unspoken yet important component of these programs was that African American patriotism informed the burgeoning civil rights activism of the war years. Home demonstration agent Fannie Boone implicitly underscored this with a talk in 1942 titled "Cooking for Freedom," which quite likely meant freedom from racialized assumptions and prejudice.[70]

African American home demonstration agents' goal to improve nutritional literacy and food security among blacks in rural Arkansas assumed particular importance during World War II. Boone's talk more practically encouraged rural blacks to support food conservation efforts because of the shortage of rationed goods during the war years.[71] This initiative had a statewide impact. In 1944, for example, Poinsett County agent Lena H. Eddington, a graduate of Arkansas AM & N and the University of Minnesota, who had begun her career as a home demonstration agent in 1941, helped black families implement dietary plans and instructed them in food production techniques. Eddington also linked good nutrition and health to national wartime goals. She also supported Live-at-Home, a program established in Arkansas in the 1920s to promote crop diversification for improved food security and self-sufficiency among farming families in poverty-stricken rural areas.[72] For black Arkansans the campaign served additional purposes. Not only were they able to improve their standard of living by becoming self-supporting, but home demonstration agents also used the program to dissuade African Americans from migrating north.[73]

Black women did not passively accept the rhetoric of the Live-at-Home program. In some instances they understood well the ways in which they could benefit from the program and further their interest in domestic reform to complement their own goals. Using local programs to improve their economic destiny and domestic environment was the most effective way for blacks to challenge and protest oppressive conditions in the rural South. The Live-at-Home program clearly helped farm woman Alwilda B. Jackson in Lexa (Phillips County). Jackson survived the Depression, combatted food insecurity, and developed her entrepreneurial and budgeting talents when she turned to raising and selling chickens at a curb market to supplement her family's income. Cultivating market power in poultry often sustained women and their families through the difficult years of the 1930s and 1940s and contributed to their economic gain.[74] Jackson maintained a meticulous "egg record," which for 1935–36 records sales of 244 dozen eggs for twenty-four cents a dozen, which amounted to $58.56 in profit. She also sold three dozen fryers for $12.60, which allowed her to feed her family and aid sick neighbors. The money Jackson earned from selling eggs allowed her to embrace better living and consumer culture: she purchased a "table radio," "kitchen linoleum," and "little odds and ends [for] around the home."[75]

In the 1950s, black agents' uplift activism did not lose momentum. Working with local home demonstration clubs, they continued to focus on home and community upkeep and food security. At a February 1950 meeting in Woodruff County, club members discussed strategies for Better Rural Living, a campaign that reflected black agents' rural activism.[76] In March 1950, Faulkner County assistant demonstration agent Maude B. Davis showed black women how to store home-cured meats and preserve eggs to increase their families' food supply. To modernize their homes, members of the Lucky Twenty home demonstration club in Lake Village in Chicot County learned to make cypress-knee lamps.[77] In 1956, in Dermott, black women learned how to use a dial telephone as part of an initiative to increase telephone availability and usage in rural areas.[78] In the same year, in Proctor (Crittenden County), the Newsome family, owners

Assistant home demonstration agents for Negro Work in 1950. Black home demonstration agents were usually well respected in rural black communities. Their uplift activism helped rural women improve their homes and communities and provided much-needed educational opportunities while carefully navigating the racial minefield of the Jim Crow South. Courtesy of Dorris Vick Collection, Manuscript Collection, 961, box 2, folder 2, image 57, University of Arkansas Libraries Special Collections.

of a fifty-three-acre farm, were able to sell $500 worth of okra, $475 worth of soybeans, $90 worth of broccoli, and, most important, $2,000 worth of cotton. Although the family was clearly successful because they diversified their crops, the skills Mrs. Newsome gained from home demonstration agents and as president of her local home demonstration club also helped her improve her family's quality of life. With agent Levada P. Mason's assistance, Newsome was not only able to produce and preserve large quantities of food for her family, she also learned—and subsequently taught other black women—how to reupholster furniture and convert kerosene lamps to electric.[79] Black agents advised farm families who owned their land to learn how to diversify their crops and increase livestock to improve their income. Rural black Arkansans who did so, they believed, were well on the way to "lifting themselves to better living by their own bootstraps."[80] Such efforts continued to underscore agents' longstanding goal of encouraging blacks to become self-sufficient producers.

African American home demonstration agents acknowledged the historical transformations in black life as the civil rights movement gained momentum following World War II. Throughout the 1950s and 1960s, State Home Demonstration Council meetings embraced such implicitly political themes as "Working Together for World Understanding," "The Homemakers' Responsibilities in Changing Times," and "Symbols of the Sixties—Alert Homemakers." These themes reflected black agents' understanding of the interconnectedness of their domestic reform activism to the changes occurring around the nation.[81]

African American home demonstration agents were critical components in rural black communities not only because they provided rural black women with home management techniques; they also often helped them obtain much-needed health-care information and advocated racial advancement by supporting the National Negro Health Movement. Although black women had long engaged in public health campaigns in their communities, the National Negro Health Movement was formally established in 1915 by Booker T. Washington. Combining the efforts of such national black organizations and institutions and based at Tuskegee Institute, the National Negro Health Movement brought national attention to African Americans' unequal health care, disproportionately high mortality rates, and poor living conditions.[82] The campaign, often understood as a movement for "race betterment," was a cooperative venture that involved the Jeanes Foundation, the National Tuberculosis Association, local interracial commissions, the USDA extension service, the American Red Cross, and the National Medical Association.[83] The movement later became National Negro Health Week and, beginning in the 1930s, was taken over by the U.S. Public Health Service.[84]

During National Negro Health Week, held in April, black home demonstration agents, educators, and leaders encouraged African Americans to participate in cleanup campaigns, attend health education programs, and obtain medical examinations.[85] In Mississippi County in 1930, 150 African Americans from Osceola, Grider, and Blytheville were examined by black doctors at clinics sponsored by the county health unit. The doctors' efforts were supported by Jeanes supervising teacher Annie Currie.[86] During World War II, local organizations such as the Parent Teacher Association, the Pulaski County Tuberculosis Association, the Negro Advisory Committee of the County Defense Council, and the Urban League cooperated with the city health department to address health concerns in black communities. Connections between improved health and the war effort were not lost on black leaders.[87] In 1943, the National Negro Health Week theme in Little Rock was "Health on the Homefront and Victory on the War Front."[88]

Although National Negro Health Week ended in 1950, black leaders continued to emphasize health concerns among African Americans. Home demonstration clubs and agents in particular continued to stress the importance of health education in rural black communities in the following years. At its seventeenth annual meeting, for instance, the state Home Demonstration Council featured a "hearing-vision" consultant from the Arkansas Department of Health who gave a talk titled "Eyes for Tomorrow." As education was always a critical pursuit for rural black women, this talk was followed with "Proper Lighting for Reading," a presentation by home demonstration agent Carreather Banks.[89] In 1954, the Burnt Hill Home Demonstration Club in Union County displayed a fair exhibit titled "Who's Ruining Your Health?" and outlined such causes of poor health as open wells and holes in house walls and screens that led to insect infestations. The exhibit also explained how electric pumps with a closed-in tank and pasteurized milk were important to improving health in rural communities. Health concerns also included mental health. The Union County Home Demonstration Council's 1954 exhibit, "Mental Health Is Everybody's Business," addressed mental illness in rural communities.[90]

African Americans were often adamant about retaining agents in their counties because they provided survival techniques. They understood that improved health was directly connected to rural development, and therefore they embraced opportunities to address and eradicate issues in their communities.[91] In 1960 in Amy (near Camden), local blacks worked with home demonstration agent Clothilde M. Shivers and the Ouachita County Rural Development advisory committee to eradicate food shortages, low incomes, and inferior medical care. Shivers, the county agent since 1952, taught rural black women food production and conservation techniques. She conducted a cooking school, giving nutritional literacy lessons to counteract dietary deficiencies. To increase black families' economic gains Shivers encouraged them to grow gardens and market their surplus produce.[92] Shivers also regularly taught a "better housekeeping methods" course at the community center near the Ivory Heights housing development and, along with the local housing authority, sponsored a Housekeeper of the Month contest.[93]

Shivers and the county advisory committee also helped Amy residents secure much-needed medical care.[94] At a time when many African Americans only had access to inferior medical care and could not afford the necessary vaccinations to avoid communicable diseases such as tuberculosis, a county nurse was enlisted to hold immunization clinics.[95] Agents further combatted the myriad health issues that plagued rural black communities by demonstrating treatments for mastitis and by highlighting the importance of mental health and dental care.[96]

Even as black demonstration agents worked to improve the lives of African Americans in Arkansas, they were impacted by the passage of such important legislation as the Civil Rights Act of 1964, which prohibited discrimination based on race, color, religion, or national origin and denied federal funds to organizations that failed to integrate. Because the Arkansas Agricultural Extension Service received federal dollars, it was forced to integrate its ranks. In 1965, the all-white Arkansas Council of Home Demonstration Clubs merged with the all-black State Home Demonstration Council. And, throughout the state, black and white county councils elected to combine their membership and make their resources available to all Arkansans.[97]

While the passage of important civil rights legislation ended segregation in home demonstration clubs at the state and local level, most rural blacks in Arkansas remained economically and politically marginalized. African American agents and the people they served still received less financial support and training from the Arkansas Agricultural Extension Service than their white counterparts. Salaries remained unequal as well. In 1965, the Union County quorum court (the county's legislative body) allocated $3,250 for the year for black county farm and home agents compared to the $8,550 set aside for white county farm and home agents.[98] Even as late as 1966, the white home agent in Hope (Hempstead County) was paid $1,500 compared to the black agent's $1,100 salary.[99] The extent to which black agents were fully integrated into the Arkansas Agricultural Extension Service is questionable as well. Black agents were often demoted and still others must have wondered if the end of the segregation negated their effectiveness in predominantly African American communities.

Although the Arkansas Agricultural Extension Service desegregated, by the 1970s only about 1.8 percent of homemaker or home demonstration clubs and only twenty out of 1,128 4–H clubs in Arkansas were actually integrated. African American agents were absorbed into the extension structure, frequently without supervisory duties.[100] When asked about desegregation, one local retired African American home demonstration agent said simply, "It was hell!"[101] Thus, ironically, desegregation had some unintended negative consequences for rural black women and the black agents who partnered with them in Arkansas.

Arkansas black home demonstration agents performed critical roles in rural communities in a period when African Americans' labor was valued more than their humanity. From agents they learned important lessons about food security, domestic economies, and community upkeep at a time when there were woefully few other resources available to effect change in rural black communities. While the philosophy of self-help, rural reform, and uplift often informed activism among African Americans, the agents most often engaged community members as partners rather than merely acting on their behalf. In doing this, black agents were aware that they must take cautious and measured steps as they navigated in and around environments in which rural blacks possessed little political or economic power. They further recognized that because they were Arkansas Agricultural Extension Service employees, it would be unwise to risk their already severely limited resources, which came in part from state and local funds, by destabilizing local racial and economic structures. By operating—through the State Home Demonstration Council in particular—within the segregated confines of the extension service, African American home demonstration agents helped rural black Arkansans, particularly women, develop leadership skills and enact domestic reform in their communities that would ameliorate the impoverished conditions rural blacks endured.

9

"Forever Lunching"

Food, Power, and Politics in Rural Ontario Women's Organizations

LINDA M. AMBROSE

On June 8, 1942, a wartime editorial on the women's page of the *Sudbury Star* newspaper in Northern Ontario reported that local women were adjusting their food intake as an act of patriotism. "With the outbreak of war, the ladies cut down considerably on the quantity and quality of these post-meeting tea parties." The writer was convinced that "this step will, more than likely, be as beneficial to the women themselves as it will be to Canada's war effort. Women, the homemakers of the nation in particular, do too much eating at any rate. They are forever lunching—if not at meetings, at afternoon teas and the like."[1] One could read the Sudbury editorialist's sarcastic insinuation that women did "too much eating" as condescending, stereotypical, and perhaps even misogynist in tone. And yet there is truth to the claim that women's groups typically paid great attention to food whether it was wartime or not. Indeed, very often, the culture of twentieth-century rural women's groups across North America, the United Kingdom, and Western Europe did revolve around food. The real question of interest is, How did women regard these activities, and what deeper meanings or purposes might lie below the surface of their preoccupation with food?

The material for this chapter is drawn from my study of the history of the Women's Institutes in Ontario; the vignettes, however, undoubtedly have many parallels among other women's organizations across North America in the first half of the twentieth century. Readers in the United States will recognize that rural Canada's "Women's Institutes" (WI) have much in common with state-sponsored Homemakers' Clubs and the educational activities provided to communities through the extension departments of state agricultural colleges, such as the ones in Arkansas that Cherisse Jones-Branch explores in chapter 8. In other parts of the world, state-sponsored groups for rural women were ubiquitous, including those across the Netherlands, as Margreet van der Burg discusses in chapter 3, and of course, the British WI, which celebrated its centenary in 2015. Women's groups in rural Ontario and elsewhere have traditionally fulfilled a threefold purpose including social roles, educational instruction, and community work. Since the end of the nineteenth century, when women's organizations burst onto the social scene to channel the energies of reformers and domestic crusaders alike, food has been central to their gatherings. At meetings sponsored by groups such as the WI, women socialized over food, studied about food preparation, and served food to their communities.

Interest in food studies has spawned many exciting new publications, but perhaps the most often cited collection in Canadian scholarship is Franca Iacovetta, Valerie J. Korinek, and Marlene Epp's edited collection, *Edible Histories, Cultural Politics: Toward a Canadian Food History*.[2] As we continue to theorize about women and food in new ways, we need to think about how women themselves perceived food-related activities and the meanings they attached to them.

On the task of building theories around women and food, Rozsika Parker's writing on needlework provides a model for this revisionist task.[3] Parker has shown that our initial interpretations of "traditionally female handiwork" are not simply what meet the eye. "The art of embroidery has been the means of educating women into the feminine ideal, and of proving that they have attained it," Parker writes. "But it has also provided a weapon of resistance to the constraints of femininity."[4] According to Parker, "women have nevertheless sewn a subversive stitch—managed to make meanings of their own in the very medium intended to inculcate self-effacement."[5] In this chapter, I will argue in a similar way that we need to look more carefully at the history of women and food to discover the ways in which women have used food to accomplish ends that are quite different from what we may have assumed. How did women use food to resist the constraints of femininity or to subversively map out new meanings of their own when casual observers might have only assumed, as the Sudbury editorialist did, that women were "forever lunching"?

Studying rural women's organizations through an examination of the records left by each local group, one is struck by how virtually every monthly minute book entry ends by stating that "a lovely lunch was enjoyed by all" or that "the hostess served a dainty cup of tea." What is a researcher to make of this shared experience that was repeated every time the women met? Because it fell outside the parameters of the business portion of the meeting, no transcript of the ensuing discussions was recorded. One can only speculate about the conversations that were exchanged over the teacups. Was the lunch served simply a social nicety that heightened the group's sense of its own prestige and led to exclusive practices? The minute books do sometimes record that limits had to be placed on the menu so that members would offer only sandwiches and one type of cake, but not pie. Evidently, some members had begun using the occasion of the monthly luncheon to demonstrate their families' economic status. It seems that some competition was brewing over who could serve the most elaborate treats, and the members had to continually call themselves back to simplicity. In addition, demonstrating one's culinary skill was part of the female culture of the day, and the monthly meeting became a site for exhibiting that talent. On more formal occasions, the minute book might record the names of those who had the honor of "pouring" at an afternoon tea. This duty often fell to young unmarried women for whom it was an occasion of both privilege and pressure.

The purpose of the lunch, members reminded one another, was not to make a burden of the task of hosting but rather to provide a forum to encourage friendships among the members. For many of the rural women who joined the Women's Institute, the monthly meeting, followed by a cup of tea, was an important antidote to the isolation they experienced in their work routines at home or on the farm. Indeed, the opportunity to gather in an all-female setting was something they looked forward to and valued very much. The WI was particularly appreciated in the northern reaches of the province, where sparse settlement dictated that one might otherwise rarely see another woman.[6] Yet even for women living in more populated parts of the province, where isolation was not necessarily so acute, the monthly meeting was a welcome diversion from routine. Wherever the branch was located, female bonding was one of the most prized outcomes of the creation of new Women's Institutes. In countless cases, those friendships formed while food was being consumed.

When the WI movement was in its early stages, as organizers wondered about the procedures they should follow at the regular meetings, concerns over the place of food were raised. The issue of whether to serve refreshments at the monthly WI meeting was debated in 1910. When the issue came to the floor of the annual convention in Toronto, some members complained that "they cannot get a large attendance without having afternoon tea," while others were convinced that food interfered with business—they could "get along better without afternoon tea."[7] Not all of the delegates agreed. One retorted: "That is not our idea. The social part is after our meeting is over, and we have the sandwiches and cup of tea. That is just how we get our members." A member from Preston explained that some of their members walked five miles for the return trip and needed to eat before they left for home. The superintendent of the Ontario Women's Institutes, George Putnam, conceded that local arrangements would vary and that sometimes, as in the Preston case, food might be necessary. He added, however, that "we regret that a few Institutes have allowed the lunch part of it to become the most prominent and apparently most interesting feature. We trust you will not do that."[8] Evidently, many branches did emphasize refreshments, albeit for different reasons.

A shift away from food-centered meetings was nevertheless emerging. A member from Waterloo County noted in 1911 that "during the first year or so of a branch Institute, they are very much interested in the planning and preparing of foods, exchanging recipes, etc. But this is invariably replaced by other subjects."[9] Indicating that food was no longer the chief focus of her local WI group, another speaker echoed the same sentiment, pointing out that with the statistical strength of the WI movement, there was the potential to move beyond tea parties and fancy work. "We are a goodly company—ten thousand women. Let us not devote too much time to ice cream, crisp salads, and eyelet embroidery. There are greater things."[10] The greater things to which she made reference included lectures from visiting professionals such as bankers and lawyers, who spoke to the groups about how to ensure legal protection for themselves by opening joint bank accounts with their husbands and getting their names on the deed to the family farm. For some local groups, especially those in Northern Ontario, they also included taking an active role in lobbying for female suffrage.[11] Evidently, when the meetings closed with a shared cup of tea, members of the WI were sometimes thinking beyond female sociability to gender solidarity as they developed strategies to improve their legal status, financial stability, or access to political rights.

Although the monthly meetings were important occasions for socializing and becoming informed on important issues of the day, the Women's

Institute organization also gained a reputation for leadership in home economics education. "In the regular monthly meetings the members have an opportunity to pass on to each other the information gained in the school of practical experience. . . . The members often refer to the Institute as 'The Great Rural University.'"[12] The association with continuing education was not limited to those monthly meetings or tips shared from members' life experiences. The idea of the WI as a "rural university" came from its offering formal training, as "short courses" sponsored by the Province of Ontario, in food preparation and nutrition.

Historian Terry Crowley has argued that WI initiatives provide one of the earliest Canadian examples of continuing education opportunities for women: in conjunction with the provincial Department of Agriculture and the Macdonald Institute in Guelph, the organization provided the services of traveling lecturers who would visit rural communities, offering short courses in cookery, garment construction, and agricultural techniques for women.[13] This mode of course delivery arose out of the reality that the new educational facility built in Guelph in 1903 at the provincial agricultural college was failing to attract farmers' daughters in the numbers that the school administration had predicted. The logical solution seemed to be that if young women could not come to the school, then the school should go to them. The very first short courses were offered off-site by Macdonald Institute in conjunction with the WIs of Haldimand County in 1912.[14] Many rural women eagerly took up the opportunity to study without leaving home, at a cost that was affordable, and on a schedule that suited the rhythm of their unpaid work in the home or on the farm. The 1913 report to the Department of Agriculture described the typical format: "The attendance at these demonstration lecture courses varied from twelve to twenty, the time period usually occupying each afternoon for ten days."[15]

This educational opportunity enjoyed sustained popularity over the next four decades, as the statistics provided in annual reports amply demonstrate. Even during World War I, when branch members were preoccupied with war work, attendance at the short courses remained consistently high. In 1915–16, fifty-five courses were offered throughout the province with an attendance of 21,643. During the war, rationing created a market for information on food conservation, and a ten-day course, "Food Values and Cooking," was developed.[16] During the 1930s, the WI continued these courses and added some of longer duration. In 1936–37, thirty-five one-month "schools in Home Economics" were held across the province, with 1,038 women in attendance. In 1936, when it became clear that many homemakers found it impossible to attend classes on a continuous basis for several weeks, special five-day and three-day versions of the courses were created.[17] In 1952, the annual report boasted that 551 short courses, most in a one-week format, had been offered, with 14,498 women attending.[18] Obviously, the programs for formal instruction in food preparation and studies in nutrition were very popular indeed.

Some historians who have explored domestic science education for women, such as Robert Stamp, interpret this phenomenon as evidence of a scheme to "teach women their God-given place" in society.[19] Thus, education centering on food has been seen as a form of social control, limiting the career options of young women by channeling them into socially prescribed gender roles. As Margaret Kechnie's scholarship has shown, attention to the pre-1920 records left by government officials and even by WI leadership confirms that this was indeed the agenda for many of the elite.[20] Ruby Heap provides a different approach to the question in her study of the experiences of two men who worked as domestic science educators. She argues that some women made very successful careers for themselves in the academic disciplines of nutrition and dietetics, working within the university as institution builders, administrators, or dietetic professionals.[21] Clearly then, these instructors were more than mere pawns of some government scheme of social regulation.

Not only was the field of domestic science potentially empowering for the instructors, but also the women who took up these educational opportunities placed a great deal of value on what they were doing. Records representing the point of view of members taking these government-sponsored courses, whether written sources or those collected through oral history, support this view. Historian Karen Dubinsky has reminded us that feminist researchers need to beware of falling into what she called "the pit of feminist arrogance" by assuming, for example, that women in certain kinds of organizations were "obviously" marginalized.[22] When exploring groups such as the WI, feminist researchers need to beware of falling into a similar trap by prematurely concluding that if the women were studying topics related to home economics, then they were being limited, stifled, and controlled.

When women recounted their experiences in WI courses, they reported opportunities for sociability with other students, skill development in practical areas that they could immediately apply, creative outlets, and certification that validated the experience. Ivy White—a young woman from Northern Ontario who completed two years of course work through the WI system in 1927 and 1928—provides an example. Almost seventy years later, Ivy proudly displayed her certificate of achievement and remembered the instruction as a highlight of her young life.[23] Whether the educational opportunities arose through formal courses or through information presented at regular monthly meetings, the opportunity to learn was attractive to women.

A poem written by a WI member from the London, Ontario, area described the monthly meeting as an important highlight in the life of "mother," a character from the postwar era who made a monthly reassignment of the usual kitchen duties in order to attend her Institute meeting.

Mother's at the Institute

That time of the month has rolled around,
The kitchen's empty, there's no sound,
And hubby knows without dispute
She's off to the Women's Institute.
A casserole will be the winner
When Mom comes home to cook the dinner,
Something fast must be the way
When WI meets for the day.
When this day comes, there is no doubt
There's no one home, the cook is out
And hubby really doesn't mind
Somewhere a sandwich, he will find.[24]

One would be hard pressed to decide which was more highly valued by this participant, the new cooking technique or the afternoon away from her usual routines. In deference to prescribed gender roles, the poet describes her husband's reaction to her outing. She explains that not only did he also welcome the news of neighborhood gossip she would bring on her return, but, as the closing lines of the poem state, "he thinks she's pretty cute, / to belong to the Women's Institute." Her new-found knowledge does not change the heterosexual dynamic of his fond (though patronizing) declaration of her "cuteness," nor does it really challenge the predetermined ideas about appropriate gender roles. Clearly, cooking was her domain; the husband requires special coping mechanisms in her absence just to feed himself.

"A Husband's View of the Women's Institute," an earlier poetic piece written in 1915, is also a comment on gender relations within the homes of Women's Institute members.[25] The poem finds the man in a similar situation, having to cope with his wife's absence while she goes off to hear "a speaker of some repute" or to deliver a paper herself to her peers at the local branch. This writer, however, fears the WI could ultimately change male and female roles.

> That my wife is busy there is no dispute,
> For she is a member of the Institute.
> I eat my mush raw and I drink cold tea
> For my wife is a busy as she can be.
> She is writing a paper on canning fruit,
> Which she will read at the Institute.
> Now I make my own bed and sweep up the floor,
> And clean all the rubbish from the door.
> . . . .
> When women get votes as I've no doubt they will,
> The first thing they'll do is to pass a new bill
> For to make it both lawful and just to shoot
> Or to herd all old men in some Institute.
> You need not expect me at home for tea
> Was the parting salute my wife gave me
> We are having a lady of some repute
> To lecture today at the Institute.

This poet/husband was concerned that women's preoccupations with food at WI were simply a front for more profound social reorganization that would assign political power to women, while men would be relegated to housework and meal preparation.

The presuffrage poet's fears about new models for the gendered division of labor in the home were never fully realized, even after women in Ontario received the right to vote in 1917.[26] Nevertheless, the association between women's meetings, food, and power is significant. For many women, access to educational opportunities (even domestic science education) was an important first step in the process of empowerment. For some, attending a WI course was their first occasion to move in a circle outside their own home and the only possible way to pursue continuing education. Regardless of their previous level of education, women found these courses granted them access to formal instruction. That alone was significant. For others, expertise in domestic science was a step toward liberation. At a very personal level, this meant that some women strategized about how to become more efficient so that they could have more leisure time. For others, leaving their homes to study food together represented a shift in the gendered division of labor while husbands learned to fend for themselves in the kitchen. Over the long term, that shift was never fully realized, of course, but just as men could learn to cook, this antisuffrage poem suggests that women would learn to vote and to use that political power to reorder the gendered construction of work in the home.

The poet's fears about complete gender reversals may have been exaggerated, but women did gain a greater voice in the public sphere through their work with food, particularly those WI members who were involved in community catering. Women's work in community food provision was most often done for the purpose of fundraising.[27] What did this work mean to the women who were involved in it? Because of their role in community food provision and the profits they generated from that work, women had an important voice in shaping community life.[28] Very often the funds on hand gave the women the financial power to provide a community meeting space.

Such was the case with the Stratton Women's Institute. "The Stratton women had a dream. It was to provide a place for the community to meet together for social and cultural growth. And so they had bake-sales, teas, suppers, anything that would raise money."[29] From the proceeds of their public food service, they bought a hall, and according to a 1929 report in the local Tweedsmuir history book, "the Hall has been the centre of social and cultural life of Stratton. Local Dramatic Club plays were highlights for many years. It has been used as a polling booth in elections. The Institute sponsored a TB Clinic."[30] In Ayr the WI branch decided to purchase a piano for the community hall. They could then dictate who would use that instrument, for what purpose, and at what price. In effect, control over their fundraising dollars allowed members to shape the nature of community leisure organization. Certainly, more was going on than compliance

In 1928 the Ilderton Women's Institute in Ontario raised money by selling meals at a Plowman's Association event. Although casual observers sometimes satirized the rural women's groups for their obsession with tea parties and cakes, women used groups such as the Women's Institute as a forum for strategic activities, including fundraising for community infrastructure and lobbying on political questions. XR1 MS A709001, Federated Women's Institutes of Ontario. Courtesy of Archival Collections, University of Guelph Library.

with the assumption that whenever a public event was hosted, the women could be counted on to make the sandwiches. Although they probably did provide the lunch, by doing so, women were generating money that gave them license to control what the next community event might be.

The WI's 1924 annual report gives a typical example of women who directed their fundraising proceeds to the local public library and cemetery. After donating to those projects, they invested in graveling the schoolyard. "This yard had been in a very muddy condition and when the Institute, through the woman school trustee, brought this to the attention of the board, the board agreed to pay for the gravel if the Institute would bear the expense of hauling it—an interesting piece of community cooperation."[31] The organization's work with schools, however, usually went far beyond improvements to the schoolyard.

As an extension of domestic responsibility in providing for one's own family, women took on the responsibility of serving as community nutritionists when they decided to provide solid nutrition for the children of the community through school lunch programs from the 1920s onward. In her study of the Women's Institutes of British Columbia, Carol Dennison has argued that members extended their homemaking skill to the community in practical ways to achieve reform.[32] In Ontario Institutes, women began by providing pots of soup, a simple act that soon gave way to political lobbying for public health campaigns and health inspections in the schools. WI members saw the provision of hot lunches as a temporary means to press the elected officials into assuming greater responsibility for the program through changed public policy. By extending their household duties these women were able to shift the traditional gendered definitions of work, stepping beyond their "female" responsibility of food provision into the previously "male" realm of politics by standing for election in their own names.

"They say the women should get on the School Boards. How are we to get there?" Mrs. W. Buchanan, a member from the Ravenna Women's Institute in Grey County, asked her audience at the Central Ontario WI Convention in 1915. "I tell you, ladies, somebody has got to go first, and it takes a lot of nerve. I have never been to a school board meeting yet, but, ladies, I intend to go." With that bold declaration, Buchanan urged her fellow members to do the same: "I hope a number of you will go to these school board meetings, because somebody has got to take hold of the work. We have heard a great deal about the work we have done, but we haven't heard about the power behind the work. I think there must be a great many trained women connected with the Institutes, or the work that has been done could not have been accomplished."[33] Providing a pot of soup for the school was only the beginning. Communal food provision soon gave way to political activism.

The minutes of women's organization meetings can speak volumes with their simple statements about women and food. As historians interested in women's experiences, and especially in the meanings that women constructed for themselves around those experiences, we must be careful to look more closely at the records that are left to us. For example, what exactly might it mean when a minute book simply reports that "the members then enjoyed a lovely cup of tea." Chances are that there was more going on than just a group of ladies "having a little tea party." Similarly, when the records show that "the members voted to bring a short course instructor on cookery," what are we to make of it? In many cases, it was more than just a cooking class designed to "teach women their God-given place," because for some of the students it may in fact have represented a chance for continuing education and an early step toward political awareness and participation. In the same way, when a community event was scheduled, the unspoken assumption was that "the members of the Institute [were] to provide refreshments." This was much more than just a shared community expectation that "the ladies could be counted on to provide." It was often a chance for those women to make money. That money meant that the women had an opportunity to determine the social organization of the community by controlling the use of the community hall. For some women, their success in catering school lunches gave them the collective political confidence to elect one of their own as an official to speak for them in the arena of municipal politics.

At one level, the minutes of an organization such as the Women's Institutes are tedious because the records of the proceedings are so repetitive. The entries, however, cryptically disguise the multiple social meanings that members of women's organizations attached to food in the first half of the twentieth century. Careful analyses of the records reveal that while it was invariably the ladies who provided the food, the food represented various meanings for the ladies. Why were women forever lunching? Perhaps it was a simple social gesture, but in some cases it was clearly their first step toward gaining access to education, community activism, or political power.

10

Frances Densmore and Mary Warren English

Indigenous Knowledge, Cross-Cultural Collaboration, and the Politics of Food

JOAN M. JENSEN

On July 25, 1917, ethnomusicologist Frances Densmore wrote to Frederick H. Hodge, chief of the Bureau of American Ethnology at the Smithsonian Institution in Washington, DC, about her work on the White Earth Indian Reservation in Minnesota: "I have secured almost as much material on foods and their preparation as is comprised in the material submitted, but it is not complete, as the Indians are collecting food-plants for identification, and other 'recipes' for cooked or combined foods are to be given me." Three years later, on July 1, 1921, Densmore wrote to Hodge's successor, J. W. Fewkes, about publishing her study on the Chippewa's knowledge of edible plants: "On my arrival here I found that Mrs. English, my faithful interpreter for Chippewa work during 14 years, has recently met with a serious accident, falling and injuring her hip. At her advanced age (almost 87 years) it is uncertain whether she will ever walk again and she is very feeble, but still able to interpret in her room, for a limited time each day. Her mind is perfectly clear and some interesting points have been added by her to the manuscript. Today we have worked with an old woman who has given interesting information on the old domestic life of the Indians. This is certainly the last opportunity for completing this book, and I hope to have it in final form by September."[1]

With these words, Frances Densmore described collecting Ojibwe knowledge about food with her long-time collaborator Mary Warren English. Densmore was overly optimistic about completing the book. English did not live to see the book because seven more years passed

Mary Warren English, principal teacher at White Earth Indian Reservation, documented Ojibwe plants and their uses with anthropologist Frances Densmore in the 1920s. She posed in this photograph, probably for Ojibwe photographer Robert G. Beaulieu, between 1910 and 1920. Courtesy Minnesota Historical Society, no. 10464040.

before its publication in 1928. In 1925, at ninety years of age, English passed away at White Earth in the home of her nephew. In the foreword to *The Uses of Plants by the Chippewa Indians,* Densmore credits the work of English and the other women who contributed their knowledge for the book: "A majority of the informants on this subject were women, and they became interested in describing the former methods of preparing vegetable foods. . . . The collection of the material herewith presented would have been impossible without the cooperation of members of the Chippewa tribe. Their assistance is gratefully acknowledged, especially that of the principal interpreter, Mrs. Mary Warren English, of White Earth, Minn., which began in 1907 and continued about 15 years."[2]

Today, *The Uses of Plants by the Chippewa* has become a classic in food studies. In the introduction to a recent reprint from the Minnesota Historical Society Press, Brenda J. Child, professor of American Studies and a member of the Red Lake Ojibwe Nation, wrote that Densmore had learned a great deal about plants from the Ojibwe, but Child does not discuss the process of that exchange. This essay explains why and how that exchange took place. Each woman brought to their exchange at White Earth a personal and family history inherited from the late nineteenth century, but each operated within shifting government food policies regarding the indigenous communities over which government officials had claimed control. The Bureau of American Ethnology (BAE) supported the research on which these two women collaborated and maintains records of their work in its archives. The legacy of their exchange seems particularly relevant for the current political issues surrounding indigenous knowledge of food and cultural sovereignty. Bringing the actors to the table of food history offers a full plate, but one that, I hope, will entice readers to further sample the research available on the history of women and food and the political context of that history.[3]

The Politics of Food

Respect and valuation of traditional knowledge of indigenous ecosystems has grown impressively in the past few decades. Today, the contest over who "owns" or has access to so-called natural resources is politically important as native tribes assert their sovereignty and treaty rights over land and its resources. Multinational corporations, however, have been busy attempting to own or control those resources—not just the physical resources but also patent rights to the genetics of plants and methods used by native tribes. Older historical discussions of how the exchange of plant knowledge took place often refer to "the Columbian exchange" or a peaceful "botanical frontier." As late as 1992, James L. Reveal called the botanical discovery of North America a "gentle conquest." A decade later, the discussion had shifted dramatically, often emphasizing conflict between native groups and a "colonizing West." Peter Zakrzeswki has rephrased the question as "bioprospecting or biopiracy?"[4]

When we look at the details of the exchange of plant knowledge, the answer to the question—prospecting or piracy—is not quite so clear. By the 1940s, the U.S. Department of Agriculture had acknowledged the importance of native plant knowledge to American agriculture as a whole. "The white man has not reduced to cultivation a single important staple during the 400 years that he has dominated the New World," wrote Everett E. Edwards and Wayne E. Rasmussen in their 1942 USDA *A Bibliography on the Agriculture of the American Indians*. During the half century preceding publication of the bibliography, scores of scholars had been documenting native plant knowledge.[5] As the present essay shows, the Densmore and English work was an important part of that transfer of knowledge, a field of study for which nonnative scholars coined the term "ethnobotany."

As ethnobotanist Nancy Turner has explained, today there is a broad movement for recognition and appreciation of indigenous peoples as knowledgeable stewards of most of the world's biodiversity. Fair and appropriate compensation of indigenous peoples for their knowledge, continued collaboration, consultation, and fair exchange of knowledge are essential components of today's ethnobotanical work. Still, we know relatively little about how that transfer of knowledge took place in the past, especially in the American West, and how it continued into the twentieth century with the support of the BAE and the USDA.[6]

Food as a Weapon of War and Peace

To understand how the attitude of the United States government toward Native American agriculture changed at the end of the nineteenth century, one has to remember the earlier politics and policy of the U.S. government in its conflicts with indigenous communities. The August 1779 destruction of Seneca food by George Washington and his troops was not the first use of the destruction of food as a weapon of war, but it is one of the best documented. By army estimates, soldiers uprooted, girdled, or chopped down 1,500 orchard trees and destroyed 60,000 bushels of corn, 2,000 to 3,000 bushels of beans, cucumbers, watermelons, and pumpkins. In all they destroyed five hundred acres of cultivated crops, an amount one officer recalled as "almost incredible to a civilized people."[7]

Nearly a hundred years later, in 1863, the governor of the territory of New Mexico sent Kit Carson to destroy Navajo wheat and corn and cut down fruit trees. That spring, Carson claimed the Navajos had raised about one thousand pounds of wheat, which he promptly confiscated. He later reported Navajos at Pueblo, Colorado, had planted large quantities of wheat and corn; according to his estimate his troops gathered more than seventy-five thousand pounds of wheat as fodder for their horses. The following year, when his troops entered Canyon de Chelly, they destroyed Navajo peach orchards. Many Navajo were incarcerated on the Bosque Redondo surrounding Fort Sumner in New Mexico. The Navajo prisoners tried but failed to raise their own crops there, and families faced starvation when the army did not provide food. Gradually, families started to escape their soldier captors and simply returned home, where they resumed growing food. Finally, in 1868, the military allowed all the Bosque Redondo families to return home to their gardens.[8]

During the intervening years, between 1779 and 1863, the U.S. government signed treaties with many tribes. Almost universally, those tribes insisted not only that fishing and hunting rights be reserved for the men but also that gathering rights be protected for the women. Native women continued to gather as well as raise food during this time. Gradually, however, national and state governments forced compromises that infringed on these gathering rights. The establishment of reservations further restricted the areas in which women could plant and gather. Still, women continued to collect food off the reservations as well as to raise food on them. As hunting became more difficult because of nonnative controls and the scarcity of remaining wild animal food, most tribes became more dependent on women increasing the supply of plant food and on men joining them in gathering it.[9]

Despite the gendered role of food production, the practice among native communities had been to share food among their members. Women were usually in control of this traditional food sharing, which was often accompanied by special song and dance ceremonies. As government agents increasingly prohibited the rituals and dances, women creatively and persistently grafted older traditions on to new ways to share food. Many rituals continued to be held secretly, and when select dancing was opened to local non-Indian spectators, these gatherings usually included at least token distribution of food. Throughout the twentieth century, many communities continued to distribute some form of food to people who viewed or participated in public ceremonies.[10]

Although treaties often guaranteed that the United States supply food subsidies to native nations, the story of the misappropriation of funds and stores is legendary. Less well known is the way agents also used food to coerce compliance with their policies. By the 1870s, the military had disarmed many of the insurgent tribes so that the ability of men to hunt was severely restricted. The government supplied some stores but expected men to work in the production of food and women to reduce their food gathering and growing. Agents sometimes provided meat and flour, and many native men joined in food production, but women continued to gather as well as grow food and to preserve and distribute it. In practice, agents also used "pinching" (withholding) food supplies to coerce women into compliance with their orders.[11]

It was no coincidence that the government started studying native food practices at the same time that they were using food to coerce the transition of indigenous communities to nonnative agriculture. Missionaries and the military had written sporadically about indigenous foodways. So too had individuals who informally exchanged information on seeds and the growing of food. Crops such as tobacco certainly interested early colonists for their potential trade with Europe, but so too did indigenous food practices. Still, the interest of the government in native plants remained scattered and informal. Early U.S. presidents all had elaborate private gardens and supported the establishment of a national botanical garden. Thomas Jefferson had even ordered Meriwether Lewis and William Clark to collect botanical information on their trek west in the early 1800s. These explorers carried with them the two best botanical books of the time: one by Carl Linnaeus, the Swedish scholar who worked out the first taxonomy for plants; the other by John Bartram, the noted American Quaker expert on plants.[12] Lewis and Clark collected very little information on how native peoples used plants. Their detailed recordings of the trek list only six brief comments about plant uses, occasionally referring in an offhand way to Indians eating some raw sprout or boiling a particular plant in their soups.[13]

The government concentrated on the plants themselves and exhibited more interest in exotic than domestic plants. The U.S. Exploring Expedition of 1838–40 collected fifty thousand specimens of foreign plants and 250 live plants according to Anne-Catherine Fallen's *A Botanic Garden for the Nation: The United States Botanic Garden*. When Congress finally established the United States Botanic Gardens in 1856, its keepers took the Royal Botanic Gardens at Kew in London as their model. Kew-trained William R. Smith, who directed the Botanic Gardens in Washington, DC, from 1853 to 1912, continued the earlier Kew tradition of showcasing exotic foreign plants.[14]

During most of the nineteenth century, then, the U.S. government seemed uninterested in how indigenous people used plants for food or for any other purpose. Government-sponsored surveys of the West displayed little more interest than did either Lewis and Clark or William Smith. In the 1850s, for example, when reporting on his boundary survey expedition, John Russell Bartlett made the offhand comment that although Southwestern tribes liked piñon nuts, the nuts were troublesome to eat. He did report on Mexican agricultural practices in southern New Mexico, however, and noted the region as a good place to collect botanical specimens. Well into the 1870s, then, the main question for government and settler alike was whether foods introduced from Europe would thrive in areas of the United States as the government expanded its reach westward across the continent.[15]

One can glimpse an inkling of change in the 1869 western explorations of John Wesley Powell. He and his wife Emma Dean Powell, who accompanied him, were both deeply interested in botany. George Vasey, who had helped Powell organize the Illinois Natural History Society, also went along on those explorations. Vasey became curator at the Smithsonian in 1872 and Powell became director of the newly formed Bureau of American Ethnology (BAE) soon after. During these early days of the BAE, it sponsored Frank Hamilton Cushing's first studies of Zuni "breadstuff" and James Mooney's studies of the Cherokee use of plant medicines. Neither of these studies was widely available in print for many years.[16]

The archaeological expeditions of the 1880s and 1890s seem to have engendered a change in attitude among both experts and the public. Gradually, the U.S. military quelled the last open resistance against its control and established the reservation system, concentrating different tribes in specified areas, mainly in the West. The civilian Bureau of Indian Affairs (BIA) took over managing these Western concentrations of native peoples. In the East, however, the BAE sponsored archeological explorations that included a search for food specimens, and in 1870 the USDA published archeologist Edward Palmer's "Food Products of the North American Indians." Palmer had played an important role in the BAE's Division of Mound Exploration, which definitively proved that the impressive earthwork mounds in the eastern United States, traditionally attributed to some group of mysterious aliens, had been created by the ancestors of indigenous people still living in the United States. During the next twenty years, as the field of ethnobotany took form, Powell and his successors in the BAE expanded their study of native plants. So too did the USDA. Unfortunately, confirmation of Native American ability to grow and gather large quantities of food was often accompanied by the Eurocentric bias of the popular conception of the evolution of civilizations, which allowed the public at large to accept previous achievements of Native Americans while still clinging to the idea that nonnatives had evolved superior contemporary cultures.[17]

The early archeologists were particularly interested in the corn that turned up in middens everywhere and appeared to be the earliest example of cultivated food crops. The government displayed many of these corn specimens at the 1893 World's Fair, and these carefully preserved examples of food eventually ended up at the University of Pennsylvania, where John Harshberger had been studying similar specimens. Harshberger had not explored native uses of food plants, but he was an avid supporter of creating gardens of native plants to illustrate their importance. His short 1893 article, "The Purposes of Ethno-botany," won him a place in botanical history as the person who named this ongoing movement to study native plants.[18]

The first scholarly academic books on ethnobotany emerged not at the University of Pennsylvania, however, but in the Midwest. In 1900, students at the University of Chicago and the University of Wisconsin produced two important scholarly studies of native plants. When Edward Barrows published his Chicago dissertation, *The Ethno-Botany of the Coahuilla Indians of Southern California,* he could point to a growing list of studies on native plants. The BAE published Albert Jenks's Wisconsin dissertation on Ojibwe wild rice as soon as he finished it. Both Barrows and Jenks easily found employment with the newly imposed colonial government in the Philippines. By 1904, the government had assigned Jenks to arrange its exhibition on the Philippines for the St. Louis World's Fair. Thus, overseas expansion kindled an increased interest in indigenous plant knowledge. Jenks proved to be an important influence on Densmore, who met him in 1904 at the fair.[19]

Early in what became a lifelong career recording indigenous music, Densmore had lectured on Indian music at the fair. Public lecture tours on topics of popular interest, often sponsored by women's clubs, had become a fairly lucrative method of earning a living for a number of middle-class women. Still barred from most higher education, women flocked to these lectures, sponsored in winter by women's clubs and some colleges and in summer through Chautauqua programs. Jenks encouraged Densmore to continue her lectures and to expand her interest in native music. He arranged for her to meet the Irogot singers whom he had brought from the Philippines to perform at the fair.[20]

Jenks and Barrows were among the few experts on native food at this time. Their work both reflected and fed the growing government and popular interest in Native American studies. While Jenks and Barrows obtained administrative positions in the colonial government in the Philippines and would settle into academic positions, there was little chance of either an administrative or academic position for Densmore. Few women were being hired by universities in any field, regardless of their specialized expertise. There was, however, some hope of gaining support from the BAE, which had sponsored some of the early musical research of Alice Cunningham Fletcher and even hired Matilda Stevenson as a field researcher. Of course, Stevenson had been married to James Stevenson, who headed the BAE until he died in 1888. Stevenson was to be an exception as a regularly employed field worker for the BAE. Still, the BAE support for Fletcher and Stevenson gave Densmore hope that she might find some sort of financial support from the government for her emerging interest in studying native cultures.[21]

Following the St. Louis fair, Densmore made her first extensive trip to visit the Ojibwe in Minnesota and Canada. According to her memoirs, the only recipe she collected on that trip was one for dandelion wine, entrusted to her by a Jesuit priest preaching among the Ojibwe. At that time, the field of ethnomusicology did not exist, and what would become ethnobotany was also not formally organized. Universities were beginning to establish anthropology departments, but positions in these, of course, went to men like Jenks who had some experience working with native people. Still, Jenks and several other men encouraged Densmore to persist in her studies of native music. After a few visits to native communities around the Great Lakes and recording some of their music, Densmore wrote to the BAE to enlist financial support for her recording project, and in 1907 the BAE gave her a stipend to continue her studies. While music would remain her central interest, Densmore also came to be an expert in the developing field of ethnobotany and the study of indigenous foods.[22]

There is no evidence that Densmore was particularly interested in studying indigenous food plants in 1907. Nor was the BAE, which, after publishing the work by Jenks on wild rice cultivation, focused on other

Frances Densmore, who disliked having her photograph taken, commissioned this formal studio portrait around 1916. Already known as an expert on Native American music, she lectured before clubs using recordings she had made for the Bureau of American Ethnology and exhibiting musical instruments she had collected. She probably used this photograph to publicize her lectures. Courtesy Minnesota Historical Society, no. 10813508.

aspects of native culture. Over at the USDA, however, there was increasing interest in native plants. The USDA had already hired specialists to study plants that might be profitably grown by farmers as a side crop and published several pamphlets urging farmers to do so. In 1909, it formally established its plant industry department, signaling its growing interest in growing native plants for profit.[23]

By 1910, the interest of the BAE in medicinal plants had also increased. In 1911, administrators sent Matilda Stevenson as their official field worker to Zuni Pueblo in New Mexico to study medicinal plants. Then, in 1915, the BAE encouraged Densmore to collect similar medicinal information among the Sioux. During World War I, in 1917, when Densmore urged the BAE to support her collection of plant specimens among the Ojibwe people, she suggested a scarcity of medicines usually imported from Europe might be one good reason for research support. The Ojibwe were noted for their medicinal knowledge as well as their knowledge about wild food plants. In her subsequent collecting of plant specimens, Densmore included plants used in healing as well as in creating objects and shelter, but food plants formed a major focus of that collection. Her suggestion seemed important because blockades of various countries during World War I had not only vastly expanded the demand for American food supplies abroad, but had fostered increased interest in growing and manufacturing medicines at home.[24]

European agriculture quickly recovered when the war ended in late 1918, but the USDA's and BAE's interest in native plants, including food plants, increased. Administrators in both agencies encouraged Densmore to continue her studies with Mary Warren English, and she did so until English died in 1925. While Densmore collected information on food plants in a number of communities, her work with English and the women of Ojibwe communities in Minnesota remained the most extensive. Within the larger context of the growing political importance of food production and interest in the value of native plants to both medicine and food, the Densmore-English collaboration flourished. With the increasing importance of food studies and the transfer of indigenous knowledge, their collaboration is worth examining in detail.

The White Earth Feast and Fair

When Densmore traveled north from her home in Red Wing to White Earth in 1907, she met English. Both were attending the annual gathering on June 13 commemorating the founding of the White Earth reservation in 1867. Many Anishinaabe people—the larger group located within the borders of the United States and Canada with which Ojibwe identify—attended these annual gatherings. Thousands of tourists also came to watch and listen as native performers danced and sang. In 1903, this "powwow" was already drawing an estimated ten thousand attendees, so when Densmore arrived, she was certainly not alone. Nor was this her first visit to White Earth. In 1905 she had stopped there briefly on the way to Grand Portage, a more isolated village, where she made her first systematic study of Ojibwe customs.[25]

The annual gathering at White Earth that Densmore attended was thus an occasion for many observers to sample Indian culture. The dances were not secret religious rituals but, as musicologist Michael McNally describes them, "folklorized performances" of traditional music. The Bureau of Indian Affairs had prohibited dances and music that accompanied older religious rituals but often allowed these more secular performances to develop around the turn of the century. The White Earth musical world was much more eclectic than these performances might indicate. As at other reservations that July, contemporary music could be heard as well, including some played by an Ojibwe all-male brass band. Other musical groups existed at White Earth, both religious and secular.

That day in 1907, when the two women first met, Densmore was forty years old and English was an elder of seventy-two. English was descended from a well-known and respected family. Her French-Ojibwe trading family—and then the Warren and English families—had acted as cultural brokers between the Ojibwe and the nonnative settlers for almost a century. Densmore's family, in contrast, were latecomers, settling in Red Wing in the second half of the nineteenth century. In the small river town of Red Wing, the Densmore family was considered to be a "pioneer" settler family. Densmore and English shared much in common, however. Both had studied music formally as young women, and both were Episcopalians. Three years after this first 1907 meeting, Densmore listed English as one of her principal collaborators when she published her first major study, *Chippewa Music*.[26]

By all accounts English was an extraordinary woman. She was the granddaughter of Marie and Michel Cadotte, whose marriage had united the Crane clan of the Chippewa with a French fur-trading family. One contemporary described Marie Cadotte as "a woman of great energy and tact, and force of character," who worked in the fur trade, including making expeditions into the hinterlands. Their daughter, Mary, married English trader Lyman M. Warren in 1821, and in 1835 gave birth to a daughter, also named Mary, on Madeline Island. Orphaned at twelve, Mary Warren was adopted by a Christian family who saw that she was well educated and proud of both her French-Ojibwe Cadotte ancestors and her Warren ancestry, which she could trace back to the *Mayflower*.[27]

Mary was eighteen when her older brother William died in 1853 at twenty-eight. William had spent his last years writing the *History of the Ojibway People*, which the Minnesota Historical Society finally published in 1885. In an unpublished manuscript Mary later described her older brother's determination to write and publish that first history of the Ojibwe, which focused on political and international history. William's history had not recounted the cultural life of the Ojibwe, and it was this history that his sister seemed determined to transmit to Densmore.[28]

As she grew up on Madeline Island, Mary attended a Presbyterian school and, after her mother's death in 1847, trained to be a teacher. She became an assistant in a mission school for Ojibwe children and then, in 1856, taught at the newly established government school in Odanah. In 1875 she moved on to Red Cliff, where she taught for fifteen years and married John English. Densmore wrote a long description of Mary's teaching career, including several years at White Earth, but said little else about the thirty-two years she had lived in Red Cliff before they met in 1907 and began the collaboration that lasted for almost twenty years.[29]

English brought to her collaboration with Densmore a lifetime of living and learning in a cultural borderland. This border culture, which never achieved the separate identity of the Canadian Métis, existed throughout the woodlands where French traders and Anishinaabe women founded mixed heritage families, and whose descendants then married into other European trading families. English was the third generation of one such family, which melded indigenous traditions to the nonnative cultures that surrounded them. Over the years, English became one of the remaining family elders who continued to tell the stories that explained how the

Ojibwe culture had survived despite persistent efforts of the U.S. government to control Ojibwe rights to land, hunting and gathering rights, and cultural traditions.[30]

In 1907, Densmore was just embarking on what would become a lifetime of studying Native American communities. At ninety, she was correcting the index of her last book on Native American music. In her later years, Densmore had often written of hearing, as a child, the music of Sioux singers drifting down from their village near Red Wing. Her family had settled in Red Wing in the nineteenth century, and Densmore counted among them a judge, a businessman father, and a mother active in the community, particularly through the Episcopalian church, of which Densmore remained a member during her lifetime. When Densmore died almost fifty years later, still living in her hometown of Red Wing, Minnesota, she could count in her archive hundreds of books and articles in native studies and thousands of records of native songs.[31]

Those were the days when Episcopalian bishop Henry Benjamin Whipple was denouncing government policies toward Native Americans, encouraging women parishioners to send clothing and needed supplies to Ojibwe communities, and ordaining young Ojibwe men as ministers for their own communities. While Whipple promoted the Christian religion, he also believed that the Ojibwe culture should be valued and preserved. English had joined the Episcopalian church during Whipple's term as bishop. Both English and Densmore approved of adopting non native religion and culture and encouraging native men to be ministers and women to be community leaders. Part of that leadership for both women was to help maintain native culture, especially Ojibwe music. They had both studied music formally, Mary at Beloit College in Wisconsin, Frances at Oberlin College in Ohio, where Bishop Whipple had also trained. Both believed that oral traditions alone could not explain the beauty and complexity of Ojibwe music and stories. In their view the music and the stories elders told about their culture needed to be preserved in permanent written form for later generations of native and nonnative alike. Both feared that the younger generation of native children who were attending schools taught in English would have difficulty maintaining their ancestral stories. As bookish women, they believed in the power of the written word to preserve oral traditions transmitted through song. To accompany the recorded music, they also collected musical instruments. They also came to share a belief that the accumulated wisdom about plants, gathered through generations of experimentation by native herbalists, was valuable and needed to be preserved in written form along with specimens of the plants.[32]

Collecting Plant Specimens

Thus, through their experience in gathering and preserving Ojibwe music and instruments and explaining how they were used, English and Densmore also came to share the task of collecting specimens of the plants and describing them. During the 1910s, as Densmore visited native communities to record their music, she expanded her plant collecting for the BAE. Although Densmore only commented briefly on the "healing arts" in her 1910 *Chippewa Music* volume, noting that skilled doctors used both music and herbs, within a few years she was collecting and sending plant specimens back to the BAE. On November 14, 1912, she wrote to the BAE from Fort Yates, North Dakota, that her Lakota informant, Chase the Bear, considered plants to be as sacred as his healing stones. "These little dried-up plants represent a great deal of toil!" she explained. By January 1914, she had sent the BAE twenty-two specimens, already identified by their Latin names, which the BAE forwarded to the Bureau of Plant Industry (BPI) of the USDA for information on which might have medicinal properties.[33]

By June 1917, three years later, Densmore had already started what was to become her major work of collecting plant specimens with English. She was back at the White Earth Reservation when she wrote to the BAE that she was collecting plant specimens used for food, including recipes for cooking them. After almost three weeks of collecting at White Earth, she went on to Mille Lacs Reservation to gather some plants that grew only in that sandier soil.[34] That August, Densmore sent a series of letters to the BAE explaining how the idea for a book on plants had occurred to her and English. Although now eighty-two, English was still collaborating with Densmore as translator and informant. They were camped at a canoe maker's home, photographing and describing his work, when the idea for the book occurred to them. There was a feeling among the Chippewa here, Frances wrote, that "the reservation is on the cusp of a general change" and that a number of the people were concerned about passing on their knowledge. Both White Earth and Mille Lacs were the best places to collect information on the use of plants, she wrote, because Indians still came there from other reservations to obtain medicines. The generation that sold these herbs not only believed in them but also used them. What had changed, she added, was that they no longer believed secrecy was necessary; the remedies could be "made a matter of record." During the next few weeks, Densmore and English collected the 160 plant specimens together, which eventually became the basis for *Uses of Plants by Chippewa Indians*. The two stayed on into September, consulting with local healers whom they paid from fifteen to twenty-five cents for each specimen and description of its use.[35]

The BAE had agreed to sponsor Densmore's research with English and to publish their work on plants as a separate bulletin. Still, the completion of this volume, with its complex process of plant descriptions in Ojibwe, English, and the Latin terminology for classifying plants, as well as their uses, dragged on. In July 1921, Densmore returned to White Earth for the first time since her extended work in 1917. She found, as she wrote in the letter quoted above, that the eighty-seven-year-old English still had a mind "perfectly clear," was able to interpret and to add some more material to the manuscript. "This is certainly the last opportunity for completing this book," she wrote her chief at the BAE, "and I hope to have it in final form by September."[36]

While English was still able to work with Densmore, she regretted that many of the other elders had died. In addition, the government had closed its school, and the whole reservation had fallen on hard times. Already nostalgic for those better times, Densmore reflected that her work had been done when White Earth was "in its best strength and in the height of its prosperity"; she hoped that "it will preserve the best life of the Chippewa for future generations." Densmore, always reluctant to comment on the policies of the BIA to her supervisors at the BAE, simply noted the reality of hard times and did not comment on the cause. Other historians have confirmed that this was an era of hard times for White Earth, attributing it to the distribution of communal land and dispersal of families to live on their separate allotments.[37]

Densmore, meanwhile, continued to work on the book on Ojibwe plant use. In December 1921, she noted in a letter that she was still waiting for a final report on the medicinal values of plants from the USDA so that she could page and index the volume. Then she appended a note about her sadness on hearing of the death of ethnomusicologist Natalie Curtis Burlin, whom she had met once at the St. Louis World's Fair in 1904, and with whom she often exchanged letters. Like Densmore, Burlin had been actively recording native music in the early twentieth century. Burlin's accidental death at forty-six (eight years younger than Densmore), would have spurred her to push for publication of the work she had completed with English.[38]

Native Uses of Plants

Five more years passed before Densmore sent the final manuscript on plants off to the BAE. It was now 1926. In a letter accompanying the manuscript, she explained again the attitude of Native Americans toward nature. In her experience, she wrote, native doctors did not consider nature as something to be "conquered." Instead, they considered nature a powerful friend and did not think of the forces of nature as "punishing" them for some activity. A medicine person might ask for a favor of nature, but healers did not believe in "conquering" nature. Densmore must have felt that her colleagues at the BAE still did not get it, despite all the efforts of English and other Native Americans to explain their attitude toward nature and plants. It is likely that English did read the typescript to check its accuracy, but she passed away in 1925, three years before the BAE finally published the book to which she had contributed so much.[39]

Densmore and English, together with a group of local herbalists and healers, believed that their practices should be preserved in writing for future generations. Others did not. Ojibwe opinion is still divided over the wisdom of sharing such indigenous knowledge publicly. The current exploitation of indigenous knowledge for private gain is now an important element in decisions about sharing that knowledge. While that healthy discussion continues, the fact remains that English and Densmore produced a unique and rare volume from their work. *Uses of Plants by the Chippewa Indians* remains not only the most complete descriptions of Ojibwe uses of plants, but of any Native American group at the time.

Some Ojibwe people still object to making plant knowledge public in written form. Much of this plant knowledge is still being passed on through religious ceremonies of the Midewiwin, the Grand Medicine Society. Historically, members of this society received instruction that might last as long as four years and culminate in a preliminary initiation, after which they might advance to higher degrees and become specialists in particular ailments. The Midew kept pictographic birch-bark records of ceremonies as well as herbal cures. Densmore wrote with admiration of these practitioners as poised, keen-eyed, and self-confident doctors who prepared specific medicines, sometimes selecting as many as twenty herbs from a pharmacology of 143 plants. "It is a teaching of the Midewiwin that every tree, bush, and plant has a use," she wrote.[40] Much as outsiders eager to understand native ways may applaud this attitude, some possessors of this knowledge believe that to spread this knowledge about the power of plants to the uninitiated would weaken it. Because earlier plant knowledge often came preserved within religious rites, that knowledge became sacred and subject to restrictions on those who inherited it. Nevertheless, much of this earlier plant knowledge had also been widely known and used in the seasonal gathering and daily preparation and consumption of food. English and Densmore, as well as many others within the Ojibwe community, thought the time had come to share that knowledge about plants as medicine, as material used for creating objects, and most important, as food.

Today, food remains not only a daily concern for native and nonnative women as well but is also a commodity subject to trade and politics. Questions of biopiracy—the practice of large corporations profiting from traditional knowledge—loom large in discussions of indigenous plant knowledge. Sustainability and health are domestic issues within Native American communities. Many communities have launched programs to expand the growing, preparation, and consumption of native plants. It is important to acknowledge, as the USDA did in the 1940s, that Euro-Americans had made no contributions to agriculture in the United States, that all of the food plants that had not been transferred from Europe were the result of Native American knowledge, not from Euro-Americans. That native women were the source of most of this food knowledge must also be acknowledged. How this knowledge about food was transferred to Europeans has been less well studied, but this example of how Densmore and English, daughters of first peoples and first settlers, developed a cross-cultural collaboration in the early twentieth century to preserve knowledge of plants and foodways is certainly an important one. The politics of collaboration is reassuring, but the politics of food remains subject to debate over who should benefit from the indigenous plant knowledge on which the food industry relies.

Notes

Introduction

1 Ken Abala, ed., *Routledge International Handbook of Food Studies* (Milton Park, UK: Routledge, 2013); Nathalie Cooke, ed., *What's to Eat? Entrees in Canadian Food History* (Montreal: McGill-Queen's University Press, 2009); Carole Counihan and Penny Van Esterik, eds., *Food and Culture: A Reader* (New York: Routledge, 1997); Marjorie DeVault, *Feeding the Family: The Social Organization of Caring as Gendered Work* (Chicago: University of Chicago Press, 1991); Franca Iacovetta, Valerie J. Korinek, and Marlene Epp, eds., *Edible Histories, Cultural Politics: Towards a Canadian Food History* (Toronto: University of Toronto Press, 2012); Jennifer Jensen Wallach, ed., *Dethroning the Deceitful Pork Chop: Rethinking African American Foodways from Slavery to Obama* (Fayetteville: University of Arkansas Press, 2015).

2 Martha H. Patterson, *Beyond the Gibson Girl: Reimagining the American New Woman, 1895–1915* (Urbana: University of Illinois Press, 2005); Martha H. Patterson, ed., *The American New Woman Revisited: A Reader, 1894–1930* (New Brunswick, NJ: Rutgers University Press, 2008); Carroll Smith-Rosenberg, *Disorderly Conduct: Visions of Gender in Victorian America* (New York: Knopf, 1985); Barbara Solomon, *In the Company of Educated Women: A History of Women and Higher Education in America* (New Haven, CT: Yale University Press, 1985).

3 R. W. Sandwell, *Canada's Rural Majority: Households, Environments, and Economies, 1870–1940* (Toronto: University of Toronto Press, 2016).

4 Carolyn M. Goldstein, *Creating Consumers: Home Economists in Twentieth-Century America* (Chapel Hill: University of North Carolina Press, 2012).

5 Melba Joyce Boyd, *Discarded Legacy: Politics and Poetics in the Life of Frances E. W. Harper: 1825–1911* (Detroit: Wayne State University, 1994), 148–49; Kendra T. Field, "'No Such Thing as Stand Still': Migration and Geopolitics in African American History," *Journal of American History* 102, no. 3 (December 2015): 693–718.

6 For the Midwest, see Joan M. Jensen, *Calling This Place Home: Women on the Wisconsin Frontier, 1850–1925* (St. Paul: Minnesota Historical Society, 2006); Mary Neth,

Preserving the Family Farm: Women, Community, and the Foundations of Agribusiness in the Midwest, 1900–1940 (Baltimore: Johns Hopkins University Press, 1995). For the Northeast, see Nancy Grey Osterud, *Bonds of Community: The Lives of Farm Women in Nineteenth-Century New York* (Ithaca: Cornell University Press, 1991), and *Putting the Barn before the House: Women and Family Farming in Early Twentieth-Century New York* (Ithaca: Cornell University Press, 2012).

Chapter 1

1 Ivy Pinchbeck, *Women Workers and the Industrial Revolution, 1750–1850* (1930; repr. London: Virago, 1981), 29–33; Frances Richardson, "Women Farmers of Snowdonia, 1750–1900," *Rural History* 11 (2014): 161–81; Nicola Verdon, "The 'Lady Farmer': Gender, Widowhood and Farming in Victorian England," in *The Farmer in England, 1650–1980*, (Farnham: Ashgate, 2013), 241–62.

2 J. Bourke, "Women and Poultry in Ireland, 1891–1914," *Irish Historical Studies* 25, no. 99 (May 1987): 293–310; Joyce Burnette, "Labourers at the Oakes: Changes in the Demand for Female Day-Laborers at a Farm near Sheffield during the Agricultural Revolution," *Journal of Economic History* 39, no. 1 (1999): 41–67, "The Wages and Employment of Female Day-Labourers in Agriculture, 1740–1850," *Economic History Review* 57 (2004): 664–90, and "Married with Children: The Family Status of Female Agricultural Labourers at Two Southwestern Farms in the 1830s and 1840s," *Agricultural History Review* 55, no. 1 (2007): 75–94; T. M. Devine, "Women Workers, 1850–1914" in *Farm Servants and Labour in Scotland*, ed. T. M. Devine (Edinburgh: Donald, 1984), 98–123; Judy Gielgud, "Nineteenth-Century Farm Women in Northumberland and Cumbria: The Neglected Workforce," DPhil thesis, University of Sussex, 1992; Valerie Hall, *Women at Work, 1860–1939* (Woodbridge: Boydell and Brewer, 2013), chaps. 5 and 6; Jennie Kitteringham, "Country Work Girls in Nineteenth-Century England," in *Village Life and Labour*, ed. Raphael Samuel (London: Routledge and Kegan Paul, 1975), 75–138; Jane Long, *Conversations in Cold Rooms: Women, Work, and Poverty in Nineteenth-Century Northumberland* (Woodbridge: Boydell and Brewer, 1999), chap. 4; Celia Miller, "The Hidden Workforce: Female Fieldworkers in Gloucestershire, 1870–1901," *Southern History* 6 (1984): 139–61; Karen Sayer, *Women of the Fields: Representations of Femininity in Nineteenth Century Rural Society* (Manchester: Manchester University Press, 1995); Pamela Sharpe, "The Female Labour Market in English Agriculture during the Industrial Revolution: Expansion or Contraction?," *Agricultural History Review* 47, no. 2 (1999): 161–81; Helen V. Speechley, "Female and Child Agricultural Day Labourers in Somerset, c. 1685–1870," PhD thesis, University of Exeter, 1999; Donna J. Ulyatt, "Female Agricultural Labour on the Dixon Estate, Lincolnshire, 1801–17," *Agricultural History Review* 54, no. 1 (2006): 79–92; Nicola Verdon, *Rural Women Workers in Nineteenth-Century England: Gender, Work and Wages* (Woodbridge: Boydell and Brewer, 2002), chap. 4.

3 *All the Year Round, A Weekly Journal Conducted by Charles Dickens, with Which Is Incorporated Household Words* (London), December 1866–June 1867, 588.

4 See Sayer, *Women of the Fields*, 70–78.

5 For example, see Arthur J. Munby's book-length poem about a strong plough woman, *Dorothy: A Country Story in Elegiac Verse* (London: Kegan Paul, 1881).

6 Joanne Shattock, "Parkes [Belloc], Elizabeth Rayner [Bessie] (1829–1925)," in *Oxford Dictionary of National Biography*, http://www.oxforddnb.com/view/article/41193.

7 Ellen Jordan and Anne Bridges, "'An Unexpected Recruit to Feminism': Jessie Boucherett's 'Feminist Life' and the Importance of Being Wealthy," *Women's History Review* 15, no. 3 (2006): 385–412; Linda Walker, "Boucherett, (Emilia) Jessie (1825–1905)," in *Oxford Dictionary of National Biography*, http://www.oxforddnb.com/view/article/31982.

8 Gerry Holloway, *Women and Work in Britain Since 1840* (New York: Routledge, 2005), 11, 43–44.

9 See Candida Ann Lacey, ed., *Barbara Leigh Smith Bodichon and the Langham Place Group* (New York: Routledge and Kegan Paul, 1986).

10 Anon., *Farming for Ladies: Or, a Guide to the Poultry-Yard, the Dairy and Piggery* (London: John Murray, 1844), quoted in G. E. and K. R. Fussell, *The English Countrywoman: Her Life in Farmhouse and Fields from Tudor Times to the Victorian Age* (1953; repr. London: Bloomsbury Books, 1985), 168.

11 Krista Cowman and Louise A. Jackson, "Middle-Class Women and Professional Identity," *Women's History Review* 14, no. 2 (2005): 165–80; 166.

12 See Bourke, "Women and Poultry in Ireland, 1891–1914"; Karen Sayer, "'His Footmarks on Her Shoulders': The Place of Women within Poultry Keeping in the British Countryside, c. 1880 to c. 1980," *Agricultural History Review* 61, no. 2 (2013): 301–29.

13 Arthur Young, *The farmer's kalendar: or, A monthly directory for all sorts of country business: containing, plain instructions for performing the work of various kinds of farms . . . By an experienced farmer* (London: Robinson and Roberts; J. Knox, 1771).

14 William Marshall, *Review and Abstract of the County Reports to the Board of Agriculture from the Several Agricultural Departments of England*, vol. 3 (London: Longman, Hurst, Rees, Orme, and Brown, 1818).

15 Harriet Martineau, *Illustrations of Political Economy*, no. 3: *Brooke and Brooke Farm*, 3rd ed. (London: C. Fox, 1833), http://hdl.handle.net/2027/inu.30000118981509.

16 For a broad introduction to British agricultural history in the modern period, see G. E. Mingay, ed., *The Agrarian History of England and Wales*, vol. 6, *1750–1850* (Cambridge: Cambridge University Press, 1989); E. J. T. Collins, ed., *The Agrarian History of England and Wales*, vol. 7, *1850–1914* (pt. 1) (Cambridge: Cambridge University Press, 2000); Alun Howkins, *Reshaping Rural England: A Social History, 1850–1925* (New York: Harper Collins, 1991).

17 On bondagers, see Gielgud, "Nineteenth-Century Farm Women"; Long, *Conversations in Cold Rooms*; Hall, *Women at Work*; Dinah Iredale, *Bondagers: The History of Women Farmworkers in Northumberland and South-East Scotland* (Wooler, UK: Glendale Local History Society, 2008).

18 Nicola Verdon, "The Employment of Women and Children in Agriculture: A Reassessment of Agricultural Gangs in Nineteenth-Century Norfolk," *Agricultural History Review* 49, no. 1 (2001): 41–55.

19 Holloway, *Women and Work*, 88–89.

20 *The Englishwoman's Review of Social and Industrial Questions*, no. 3 (April 1867), 196.

21 Ibid., 195.

22 30 and 31 Vict. c. 130, An Act for the Regulation of Agricultural Gangs, August 20, 1867.

23 Lacey, *Barbara Leigh Smith Bodichon*, 1–2.

24 *Englishwoman's Review*, no. 11 (April 1869), 155–76, 157. For an overview of agricultural trade unionism, see J. P. D. Dunbabin, "The Incidence and Organisation of Agricultural Trades Unionism in the 1870s," *Agricultural History Review* 16, no. 2 (1968): 114–41.

25 *Englishwoman's Review*, no. 11 (April 1869), 165.

26 Ibid., 172.

27 Martineau (1842–1909) wrote on agricultural topics as a theme within her political economic writing. Walker, "Boucherett."

28 "Miss Boucherett, of Willingham, has established a system, which has done much good, of taking out girls from the workhouse and putting them in small places to learn domestic duties." Royal Commission on the Employment of Children, Young Persons and Women in Agriculture, British Parliamentary Papers 1867–68, 17: 85; see also Walker, "Boucherett."

29 *Englishwoman's Review*, no. 11 (April 1869), 175.

30 A full listing of the topics under the heading Employment/Agriculture within the index to the *Englishwoman's Review of Social and Industrial Questions*, digitized from A. M. Mackenzie, H. Blackburn, C. A. Biggs, E. J. Boucherett, *The Englishwoman's Review of Social and Industrial Questions* (New York: Garland, 1979), http://hdl.handle.net/2027/uc1.31158010772001?urlappend=%3Bseq=117.

31 The contact for the Association for the Promotion of Food Production by Women was Hon. Sec. Miss Thorne, Southover Grange, Lewes, Sussex. Articles and other entries on topics related to food production in 1881 include Anon., "Art V.—Dairy Farming for Women," *Englishwoman's Review*, January 15, 1881, 9–11; Anon., "Food Production," *Englishwoman's Review*, November 15, 1881, 511–12; Mrs. M. E. Parker, "Art. II.—Healthy Work for Women," *Englishwoman's Review*, July 15, 1881, 295–99.

32 Paul Brassley, "Agricultural Science and Education," in *The Agrarian History of England and Wales*, vol. 7: *1850–1914*, ed. E. J. T. Collins (2000), 633, table 8.1.

33 B. Leigh Smith Bodichon, "Reasons for the Enfranchisement of Women," a paper read at the National Association for the Promotion of Social Science, October 1866, in Lacey, *Barbara Leigh Smith Bodichon*, 105–7. Matilda Betham Edwards was brought up on a farm in Suffolk; when her father died she continued to manage the tenancy with her sister for two to three years, until the property was split

up. Her interest in writing then supplanted her farming—her first novel, which she had begun as a teenager, was published in 1857. After losing the farm, she traveled Europe with Bodichon and wrote about it in *A Winter with the Swallows* (1866) and *Through Spain to the Sahara* (1867). She became a full-time travel writer, journalist, and novelist on her return. Her long-standing interest in France and French agriculture was rewarded in 1891 when she was decorated by the French government. See Helen C. Black, *Notable Women Authors of the Day* (London: MacLaren, 1906), 120–30; Janet Todd, ed., *Dictionary of British Women Writers* (London: Routledge, 1989), 62–63; Claire Buck, ed., *Bloomsbury Guide to Women's Literature* (London: Bloomsbury, 1992), 341; Joan Thirsk, "Matilda Betham Edwards (1836–1919)," *Oxford Dictionary of National Biography* (Oxford: Oxford University Press, 2012).

34 Leigh Smith Bodichon, "Reasons," 106–7.

35 *Census of England and Wales, 1901: General Report with Appendices*, British Parliamentary Papers (London, 1904), 103. On widow farmers, see Verdon, "The 'Lady Farmer,'" 241–62.

36 *Englishwoman's Review*, no. 1 (January 1870): 11–12; no. 5 (January 1871): 66–67.

37 *Englishwoman's Review*, no. 14 (April 1873): 156; no. 18 (April 1874): 87, 144; no. 52 (August 1877): 376; no. 53 (September 1877): 426.

38 *Englishwoman's Review*, no. 69 (November 1879): 481.

39 Ibid., 484.

40 *Englishwoman's Review*, no. 82 (February 1880): 85; and no. 83 (March 1880): 101–2, 116.

41 Janet Horowitz Murray and Anna K. Clark, *Englishwoman's Review of Social and Industrial Questions: An Index* (New York: Garland, 1985). See July 1880, no. 87, 332–33, an extract from Richard Jefferies, *Hodge and His Masters*, on farmer's daughters; August 1880, no. 88, 373, on new opening for agricultural education; January 1881, no. 93, 9–11 on the new dairy schools; September 1883, no. 125, 400–404, a review of Mrs. J. H. Burton, *My Home Farm*; September 1884, no. 137, on 434 prizes awarded for dairy work, cheese making, calf rearing, etc., at the Model Farm, Glasnevin, Ireland; October 1888, no. 185, 472–73, "Lady Farmers"; October 1898, no. 239, 282, use of article by Mrs. Virginia M. Crawford on "Englishwomen and Agriculture" in *Contemporary Review*; January 1901, no. 248, 51–53 report on lecture by Mrs. Chamberlain, member of the Women's Agricultural and Horticultural Union to the Women's Institute

42 Boucherett's "How to Provide for Superfluous Women" appears in Josephine Butler, ed., *Woman's Work and Woman's Culture* (London: Macmillan, 1869), 27–48.

43 Nicola Verdon, "Business and Pleasure: Middle-Class Women's Work and the Professionalization of Farming in England, 1890–1939," *Journal of British Studies* 51 (April 2012): 393–415.

44 Gillian Sutherland, *In Search of the New Woman: Middle-Class Women and Work in Britain, 1870–1914* (Cambridge: Cambridge University Press, 2015).

45 Frances, Countess of Warwick, *Life's Ebb and Flow* (London: Morrow, 1929), 250–51. On the wider back-to-the-land movement, see Jan Marsh, *Back to the Land: The Pastoral Impulse in England from 1880–1914* (London: Faber and Faber, 1982).

46 *Ladies Field*, February 17, 1906, 437.

47 Warwick, "New Women," 1–2.

48 Edith Bradley and Bertha La Mothe, *The Lighter Branches of Agriculture* (London: Chapman and Hall, 1903).

49 On the curriculum followed at Swanley, see D. Opitz, "'A Triumph of Brains over Brute': Women and Science at the Horticultural College, Swanley, 1890–1910," *Isis* 104, no. 1 (March 2013): 30–62.

50 Mrs Roland Wilkins, "The Work of Educated Women in Horticulture and Agriculture," *Journal of the Board of Agriculture* 22 (April 1915–March 1916), 622.

51 Sutherland, *In Search of the New Woman*, 166.

Chapter 2

1 The Sawtelle family has generously shared their archives with me. Historians, rural sociologists, and friends have also offered crucial suggestions. National Archives and Records Center, College Park, Maryland, RG 83, Farm Women Survey and Correspondence, box 12, entry 147, Mrs. T. J. Downs, January 22, 1921; Emily Hoag to Charles Galpin, undated early 1921; JW to Iva C. Gould, February 3, 1921. Hereafter cited as NARC, FWSC, RG 83.

2 Martha H. Patterson, *Beyond the Gibson Girl: Reimagining the American New Woman, 1895–1915* (Urbana: University of Illinois Press, 2005); Martha H. Patterson, ed., *The American New Woman Revisited: A Reader, 1894–1930* (New Brunswick, NJ: Rutgers University Press, 2008); Elizabeth Otto and Vanessa Rocco, *The New Woman International: Representations in Photography and Film from the 1870s through the 1960s* (Ann Arbor: University of Michigan Press, 2012). The term "new patriarchy" in the agricultural professions is from Rebecca S. Montgomery, "'We Are Practicable, Sensible Women': The Missouri Women Farmers' Club and the Professionalization of Agriculture," in *Women in Missouri History: In Search of Power and Influence*, ed. Lee Ann Whites, Mary C. Neth, and Gary R. Kremer (Columbia: University of Missouri Press, 2004), 199.

3 Hoag's work was cited briefly in Pitirim A. Sorokin and Carle Z. Zimmerman, *Principles of Rural-Urban Sociology* (New York: Holt, 1929), and in Carl Taylor, *Rural Sociology and Its Economic, Historical, and Psychological Aspects*, rev. ed. (New York: Harper, 1933). Julie N. Zimmerman and Olaf F. Larson, *Opening Windows onto Hidden Lives: Woman, Country Life, and Early Rural Sociological Research* (University Park: Pennsylvania State University Press, 2010), 62.

4 From census data we can track the Hoag family origins, its composition, and location as it moved from Saratoga to Washington and then Broome County, New York.

5 A detailed analysis of women employed by the USDA can be found in Records of the Office of the Secretary of Agriculture, General correspondence of the Office of the Secretary, 1906–70, RG 16, box 722, Edwin T. Meredith to Mary Holland Kinkaid, May 24, 1920, "Agriculture Department Employees (Women)." Florence E. Ward was

the best known of the USDA women employees. The USDA published her study *The Farm Woman's Problems* in November 1920, as circular 145. She was an expert on kindergarten work, especially the new Montessori methods, taught at Northern Iowa Teacher's College from 1906 to 1914, then worked for the USDA in Washington, DC. The only other woman I have been able to trace is Emma L. Reh Stevenson (1896–1965), who worked as a chemist for three years during this time before moving on in 1924 to the newly established Smithsonian Science Agency, which publicized scientific work sponsored by the government. She worked there until 1935, then conducted field research for the Soil Conservation Agency. David L. Browman, *Cultural Negotiations: The Role of Women in the Founding of Americanist Archaeology* (Lincoln: University of Nebraska Press, 2013), 7, 120.

6 On the lives of rural women in that area, see Grey Osterud, *Putting the Barn before the House: Women and Family Farming in Early Twentieth-Century New York* (Ithaca: Cornell University Press, 2012).

7 The pageant pamphlet is in the Sawtelle family archives. She satirized her educational experiences in Emily F. Hoag, "Dreams: The Duchess Comes Galumphing Back," *School and Home Education* 34 (October 1914), 75.

8 Hoag chose the University of Wisconsin over Cornell University, which was closer to home. On the conflicts between male professionals running the extension service in the College of Agriculture and the newly formed College of Home Economics, headed by Martha Van Rensselaer, see Kathleen Rebecca Babbitt, "Producers and Consumers: Women of the Countryside and Cooperative Extension Service Home Economists, New York State, 1870–1935," PhD diss., State University of New York at Binghamton, 1995; Kathleen R. Babbitt, "The Productive Farm Woman and the Extension Home Economist in New York State, 1920–1940," *Agricultural History* 67, no. 2 (Spring 1992): 83–101; and e-mail to the author, April 28, 2010. Babbitt also assisted me in locating biographical information on Hoag.

9 On the country life movement, see David B. Danbom, *The Resisted Revolution: Urban America and the Industrialization of Agriculture, 1900–1930* (Ames: Iowa State University Press, 1979), 167–72; and Mary Neth, *Preserving the Family Farm: Women, Community, and the Foundations of Agribusiness in the Midwest, 1900–1940* (Baltimore: Johns Hopkins University Press, 1995), 102–6.

10 For Wisconsin, see Charles Josiah Galpin, *Rural Life* (New York: Century, 1918); and Joan M. Jensen, *Calling This Place Home: Women on the Wisconsin Frontier, 1850–1925* (St. Paul: Minnesota Historical Society Press, 2006), 90–143.

11 See Malcolm Rutherford, *The Institutionalist Movement in American Economics, 1918–1947* (New York: Cambridge University Press, 2011), especially the section on "Wisconsin Institutionalism," 180–222.

12 NARC, FWSC, RG 83, box 17, "Teaching of Rural Sociology." Scholars interested in the Country Life Movement introduced most of the early rural sociology courses between 1905 and 1915. The University of Chicago is sometimes credited as introducing the first course in 1894; Michigan State College as offering a course in 1903; University of Massachusetts in 1907. By 1925, over half of the sixty-eight land-grant

colleges offered rural sociology courses. Seventeen of these colleges were "for colored students."

13 National Archives, U.S. Department of Agriculture, RG 80, Records of the Bureau of Agricultural Economics, Division of Farm Population and Rural Life and Its Predecessors, General Correspondence, 1919–1934, Farm-Women, box 12, entry 147, E. C. Lindeman to Galpin, October 26, 1920; Margaret F. Byington to Galpin, November 10, 1920; Howard F. Odum to Galpin, October 29, 1920; Liberty Bailey to Galpin, November 5, 1920.

14 For Galpin's early methods, see his *A Method of Making a Social Survey of a Rural Community*, University of Wisconsin AES Circular of Information 29 (Madison: University of Wisconsin: 1912), and *The Social Anatomy of an Agricultural Community, AES Research Bulletin* 34 (Madison: University of Wisconsin, 1915); "Rural Relations of the Village or Small City," Fourth Wisconsin Country Life Conference, January 1914, *Bulletin of the University of Wisconsin* 711 (Madison: University of Wisconsin, 1914); and *Rural Social Centers in Wisconsin, AES Bulletin* 223 (Madison: University of Wisconsin, 1914).

15 Amy Hague, "'What If the Power Does Lie Within Me?' Women Students at the University of Wisconsin, 1875–1900," *History of Higher Education* 4 (1984): 78–100. The professions of three graduate students, Kate Everest Levi, Louise Kellogg, and Lorian Jefferson illustrate these changes. Everest received her PhD in history under Frederick Jackson Turner in 1893, reportedly the first doctoral degree awarded by the history department and the first PhD in history awarded to a woman in United States. Her research on German immigrants was based on questionnaires that showed her debt to new sociological methods. She studied economics with Ely as well as history with Turner. After graduation, she accepted a job at a Pittsburgh settlement house, but after her marriage she resigned her position and disappeared from the public record. Kellogg received her PhD in history in 1900. She is remembered as one of Turner's most important students, who set forth the basic themes of the colonial frontier theory and obtained fellowships to study at the London School of Economics. She received a prize for her work from the American Historical Association, became what we would now call a "public" historian working at the University of Wisconsin Library, and helped shape programs that fostered education for the entire population of Wisconsin. She never married. Jefferson obtained her MA in 1907 and worked on her PhD in the College of Agriculture during the 1910s. See Anne Moore's chapter in this volume for her subsequent career. Other women were obtaining positions in the Department of Agriculture at the time. Like Kellogg and Jefferson, many of them never married.

16 Richard T. Ely and Charles J. Galpin, "Tenancy in an Ideal System of Landownership," *American Economic Review* 9 (March 1919): 180–212; reference to Hoag survey, 196. C. J. Galpin and Emily F. Hoag, *Farm Tenancy: An Analysis of the Occupancy of 500 Farms*, Agricultural Experiment Station of the University of Wisconsin, *Research Bulletin* 44 (February 1919), 3–4. For the European roots of the theory, see Joshua M. Nygren, "In Pursuit of Conservative Reform: Social Darwinism, the Agricultural

Ladder, and the Lessons of European Tenancy," *Agricultural History* 89, no. 1 (Winter 2015): 75–101.

17 On Carney's work in the United States, see Richard Glotzer, "The Career of Mabel Carney: The Study of Race and Rural Development in the United States and South Africa," *The International Journal of African Historical Studies* 29, no. 2 (1996): 309–36. On Ward, see also Charles J. Galpin, *My Drift into Rural Sociology: Memoirs of Charles Josiah Galpin* (Baton Rouge: Louisiana State University, 1938), 34–35.

18 For a glimpse into the segregated District of Columbia in the 1920s, see the fictional account of Breena Clark, *River Cross My Heart* (Boston: Little Brown, 1999).

19 Charles J. Galpin, "Farm Life Studies and Their Relation to Home Economics Work," *Journal of Home Economics* 12, no. 4 (1920): 159–61; reprinted in Zimmerman and Larson, *Opening Windows onto Hidden Lives*, 140–44.

20 Galpin moved Veda Larson (whom he later described as having "a rare feminine accomplishment," a facility with mathematics) to the Census Bureau to work with the 1920 farm census material. This job, tabulating farm population from eight widely scattered counties, took her and several clerks two years to finish. While Larson gathered data that would allow him to have a sample from the entire country to evaluate, Galpin concentrated on locating data sets that were readily available for smaller studies. Galpin arranged one of his first cooperative projects with the University of Montana at Bozeman for Walter Baumgartel, who soon went off to work there. He dispatched W. C. Nason, a graduate from the University of Minnesota whom Taylor had already assigned to work with Galpin, into the field to study rural community buildings. Galpin, *My Drift*, 35–44, discusses these first projects and the first budget fight.

21 Emily F. Hoag, *The National Influence of a Single Farm Community: A Story of the Flow into National Life of Migration from the Farms*, U.S. Department of Agriculture Bulletin no. 984 (Washington, DC, 1921); Walter H. Baumgartel, *A Social Study of Ravalli County, Montana*, AES Bulletin no. 160 (Bozeman: University of Montana, 1923); Charles J. Galpin and Veda B. Larson, *Farm Population of Selected Counties, Composition, Characteristics, and Occupations in Detail for Eight Counties, Comprising Otsego County, N.Y; Dane County, Wis.; New Madrid and Scott Counties, Mo.; Cass County, N. Dak.; Wake County, N.C.; Ellis County, Tex.; and King County, Wash.* (Washington, DC: U.S. Bureau of the Census, 1924).

22 Zimmerman and Larson, *Opening Windows*, 37n22; David F. Houston, "Report of the Secretary of Agriculture," *Yearbook of the United States Department of Agriculture, 1913* (Washington, DC: Government Printing Office, 1914), 9–74.

23 David Houston, "The American Farm Woman as She Sees Herself," *Yearbook of the Department of Agriculture, 1915* (Washington, DC: Government Printing Office, 1915), 311–18.

24 Ibid.

25 Zimmerman and Larson, *Opening Windows*, 39, and Jellison both comment about the USDA policy of not interfering with domestic arrangements of the farm household. The same policy applied to racial arrangements. See especially Carmen V.

Harris, "'The Extension Service Is Not an Integration Agency': The Idea of Race in the Cooperative Extension Service," *Agricultural History* 82, no. 2 (Spring 2008): 193–219.

26 NARC, FWSC, RG 83, Galpin to Phelps, November 20, 1920; Galpin to Herbert T. Myrick, November 30, 1920; Galpin to Edwin C. Powell, October 11, 1920; Edwin C. Powell to Galpin, December 23, 1920.

27 NARC, FWSC, RG 83, Christine N. Smith to Emily Hoag, October 12, 1920, discussing the Southern state survey; Galpin to President K. S. Butterfield, Amherst, Massachusetts, December 20, 1920, asking for names of fifteen women for a 100-woman study to start in January, 1921.

28 Janet Galligani Casey, *A New Heartland: Women, Modernity, and the Agrarian Ideal in America* (New York: Oxford University Press, 2009), 53–82, discusses the *Farmer's Wife* and its purchase by Edward Webb in 1905. She refers to Mary Meek Atkeson as an "astute" commentator (51) but does not discuss Hoag.

29 NARC, FWSC, RG 83, Bess M. Rowe to Galpin, March 15, 1922. The *Farmer's Wife* was in print from 1906 to 1939.

30 Apparently, Galpin was making plans for further visits to Tennessee, Michigan, Indiana, and Ohio. There seems to be no correspondence remaining for these visits, if they did take place. In no way were the regions systematically sampled. Hoag did draw examples in her final report from women's letters from the Midwest and East, but none from the Far West, except for a few from Idaho. It is probable that she did not interview in these areas.

31 NARC, FWSC, RG 83, Galpin to Nat T. Frame, West Virginia University, January 4, 1921; Hoag to Galpin, January 27, 1921.

32 NARC, FWSC, RG 83, Hoag to Galpin, February 8, 1921.

33 For a sample of the Harper letters describing these black women farmers, see Melba Joyce Boyd, *Discarded Legacy: Politics and Poetics in the Life of Frances E. W. Harper, 1825–1911* (Detroit: Wayne State University Press, 1994), 148–49. On South Carolina, see NARC, FWSC, RG 83, Hoag to Galpin, November 17, 1920; on the North Carolina conversation, see Hoag to Galpin, November 2, 1920. On racial attitudes, see Zimmerman and Larson, *Opening Windows*, 57.

34 NARC, FWSC, RG 83, Hoag to Galpin, February 8, 1921, on the train from West Virginia.

35 Comments on reticence and bashfulness are in NARC, FWSC, RG 83, Hoag to Galpin, March 6, 1921, and Hoag to Galpin, March 29, 1921.

36 NARC, FWSC, RG 83, Cola L. Fountain to Dwight Sanderson, February 1, 1921, and Hoag to Galpin, July 11, 1921. Sanderson himself was still fighting for the recognition of rural sociology as an academic discipline, according to Lowry Nelson, *Rural Sociology: Its Origins and Growth in the United States* (Minneapolis: University of Minnesota Press, 1967), 55–56.

37 NARC, FWSC, RG 83, Mabel G. Feint to Dwight Sanderson, January 29, 1921.

38 It is not clear just how Hoag and Galpin negotiated working during the split between Sanderson in agricultural economics and the extension service, which supported women joining the Farm Bureau and organizing separate groups there, and Van Rensselaer in home economics, who wanted separate training and organizing of

women into study clubs. See Babbitt, "Producers and Consumers," 309, and Babbitt, "The Productive Farm Woman."

39 NARC, FWSC, RG 83, Abbuhl to Galpin, June 22, 1921, and Abbuhl to Sanderson, January 1921.

40 NARC, FWSC, RG 83, Hoag to Galpin, July 11, 1921.

41 Emily Hoag Sawtelle, "The Advantages of Farm Life," 1–2. Sawtelle's entire text is reprinted in Zimmerman and Larson, *Opening Windows*, 149–85. The reprint is identical to the copy in the archives with the exception of Hoag's appendix of tables.

42 See Zimmerman and Larson, *Opening Windows*, 32–40, for an excellent summary of research on home economics and extension. U.S. Department of Agriculture, Office of the Secretary, *Report of the Committee Appointed by the Secretary of Agriculture to Consider Plan of Organization, Scope of Work, and Projects for the Office of Farm Management* (Washington, DC: Government Printing Office, 1919), circular 132; and *Report of Committee Appointed by the Secretary of Agriculture to Consider the Subject of Farm Life Studies as One of the Divisions of Research Work of the Proposed Bureau of Farm Management and Farm Economics* (Washington DC: Government Printing Office, 1919), circular 139.

43 Hoag, "The Advantages of Farm Life," 149–85.

44 Department of Agriculture, *Official Record* (Washington, DC: Government Printing Office, 1922), 203, 218, 395. Indeed, the gender hierarchy did not drastically change in the next eighty years. The first woman to head the USDA was Ann Veneman, the extremely well-qualified woman appointed by President George W. Bush; Veneman served as secretary of agriculture from January 2001 to January 2005.

45 For Hoag's talks, see U.S. Department of Agriculture, *Official Record* 1, no. 11 (March 15, 1920): 7.2, and no. 25 (June 21, 1922): 7.1. The Maine government was so desperate to retain women teachers for their more than 2,500 rural schools—from which, in 1920, more than half of the beginning teachers resigned before the end of their first year teaching—that the legislature committed one million dollars to upgrade the schools. Maine, State Commissioner of Education, *Report for School Year Ending June 31, 1920* (Waterville: Sentinel, 1921), esp. 40–45.

46 Carolyn M. Goldstein, *Creating Consumers: Home Economists in Twentieth-Century America* (Chapel Hill: University of North Carolina Press, 2012), esp. 3–11.

Chapter 3

This article is based on earlier work published in Dutch. For more nuances and details, see Margreet van der Burg, *"Geen tweede boer": Gender, landbouwmodernisering en onderwijs aan plattelandsvrouwen in Nederland, 1863–1968* ["No second farmer": Gender, agricultural modernization and education for rural women in the Netherlands, 1863–1968] (Hilversum, Netherlands: Veloren, 2002); in Dutch with extended summary in English; Margreet van der Burg, "Rural Women's Voices and the Identification of Rural Women's Issues within the Landscape of International Organisations, 1889–1940," in *Grenzregime: Geschlechterkonstellationen zwischen Kulturen und Raumen der Globalisierung* [Border regime: Gender constellations between cultures and spaces of globalization], ed. Waltraud Ernst (Berlin: Lit-Verlag, 2010), 203–27.

1 Farm household management teacher quoted in Burg, *"Geen tweede boer,"* 302, in Dutch: "Als het om ontwikkeling gaat, spreek je niet alleen over voeding, je spreekt ook over gezondheid, voeding en landbouw. Er komen dan allerlei zaken ook aan de orde zoals schoon water, hygiëne enz."

2 Compare Penny van Esterik, "Right to Food; Right to Feed; Right to Be Fed: The Intersection of Women's Rights and the Right to Food," *Agriculture and Human Values* 16 (1999): 225–32.

3 See Burg, *"Geen tweede boer"* and "Rural Women's Voices" for more examples.

4 See Burg, *"Geen tweede boer,"* chap 3.

5 See Burg, "Rural Women's Voices."

6 Paul De Vuyst, *L'enseignement agricole aux jeunes filles* [Agricultural education for girls] (Brussels, Belgium: Weissenbruch, 1891), and *De maatschappelijke rol der boerin: Hare opleiding, de boerinnenkringen, hunne inrichting in het buitenland, practische wenken* [The social role of the farm woman: Her education, farm women circles, foreign examples, and practical advice] (Brussels, Belgium: Dewit, 1907).

7 Paul De Vuyst, *Women's Place in the Rural Economy: A Study in Sociology*, trans. Nora Hunter (London: Blackie, 1913).

8 *VIII. Internationaler landwirtschaftlicher Kongress*, Wien 21.-25. Mai, 1907, Protokoll über die Beratungen der Sektion IIA Land- und forstwirtschaftlichen Unterrichtswesen [International Agricultural Congress, Vienna, May 21–25, 1907, Proceedings of sec. 2a, Agricultural and forestry education], international 620–21; discussions 215–18.

9 See also Leen Van Molle, *Ieder voor allen: De Belgische Boerenbond 1890–1990* [Each one for all: The Belgian Farmers' Union 1890–1990] (Leuven, Belgium: Economaat van de Belgische Boerenbond, 1990, 118; Burg, *"Geen tweede boer"* and "Rural Women's Voices."

10 Burg, *"Geen tweede boer,"* esp. 145–48; Burg, "Rural Women's Voices."

11 See Burg, *"Geen tweede boer,"* chap. 4.

12 *50 jaar "Huis te Lande": Tuinbouwschool voor meisjes te Rijswijk Z.H., 1907–1957* [Fifty years "Huis te Lande": Horticultural schools for girls at Rijswijk, S-Holland, 1907–1957] (n.p.: n.p.); Burg, *"Geen tweede boer,"* 52.

13 "Landbouwtentoonstelling te Roermond" [Agricultural Exhibition at Roermond] (June 16, 1909), Nieuwe Rotterdamsche Courant (NRC), June 17, 1909, Ochtendblad A, saved in Atria, Archive of Nationaal Bureau voor Vrouwenarbeid (NBV) [National Office for Women's Labour] 329, j. 504; in Dutch:

> Wie verzorgt en voedert op de boerderij de melkkoeien, neen, al het vee, groot en klein, tot de kippen toe? De vrouw. Wie praat bijgevolg nogal eens mee over het kopen van krachtvoer? De vrouw. Wie fokt de kalvers en jonge biggen op? De vrouw. Wie melkt en behandelt verder de melk? De vrouw. Wie bezorgt heel veel ongerief, als de groenten in den tuin niet lukken? De vrouw. Wie moet voor doelmatige voeding voor het gezin zorgen? De vrouw. Welnu zou het dan niet nuttig zijn, dat onze boerendochters onderricht ontvingen in: Gezondheidsleer en veevoeding? Veeteelt? Zuivelbereiding? Het bemesten en doel-

matig bewerken enz. van den tuin. Eenvoudig boekhouden? Het samenstellen van een eenvoudig, goedkoop en toch voedzaam menu?

14 A more detailed analysis of discourses and negotiations can be found in Burg, *"Geen tweede boer,"* esp. chaps. 5–7.

15 The first Association of Former Pupils (Vereeniging van Oud-Leerlingen) was set up in Kollum, province of Friesland, in 1915. In Friesland also the first provincial association was founded in 1919. See also Margreet van der Burg, *Een half miljoen boerinnen in de klas: Landbouwhuishoudonderwijs vanaf 1909* [Half a million farm women in the classroom: Farm household management education from 1909 onward] (Heerlen, Netherlands: De Voorstad, 1988).

16 Minutes of meeting, Inspector of agricultural education with farm household management teachers, December 20, 1916, Archives of (Nieuw) Rollecate. After I had completed my research, the school materials were sorted and only a selection was archived under Rijksarchief (RA) Overijssel, Zwolle, Archief Rijksscholengemeenschap (Nieuw) Rollecate (1911) 1913–1976 (1977), inv. 435.

17 Greta Smit, *Het landbouwhuishoudonderwijs in Nederland: Ontstaan, ontwikkeling en betekenis 1908/1965* [Farm household management education in the Netherlands: Start, development and significance 1908–1965] (The Hague, Netherlands: Staatsdrukkerij- en Uitgeverijbedrijf, 1966, 72; in Dutch: "Nooit werd er op een—toen nog veelal gebruikelijke—beschermende manier gesproken over arme en onontwikkelde mensen. Integendeel, wij werden opgevoed tot respect en bewondering voor de dappere vrouwen, die onder zo zware omstandigheden haar taak moesten verrichten."

18 *25 jaar landbouwhuishoudonderwijs 1913–1938* [Twenty five years of farm household management education 1913–1938] (Deventer: Davo, 1938), 24–25; in Dutch: "De huishouding als *een organisch geheel* te leren zien, een bedrijf, dat gevoerd wordt in het belang der gezinsleden, dus der *mensen*, en dat schoonmaken, koken wassen, óm het schoonmaken, koken en wassen zelf, geen zin heeft, als het niet gebeurt in het waarachtig belang der mensen."

19 *Bijblad van In en Om de Woning* 6, no. 9 (September 1919): 77, in Dutch: "Bij het leven van een vrouw op de boerderij grijpen haar taak als huisvrouw en als boerin zoo in elkaar, dat deze niet of heel moeilijk van elkaar te scheiden zijn, behalve wellicht bij heereboerderijen. Zoo is bij het l.b.h.o. het landbouwonderwijs een integreerend bestanddeel van dit onderwijs, en zal het l.b.h.o. nooit aan zijn doel beantwoorden, als dit laatste niet even goed tot zijn recht komt als het eerste."

20 *In en Om de Woning* 25, no. 9 (September 1919): 85; in Dutch: "Het protest van de leeraressen bij 't L.H.O. zal wellicht nogmaals met succes overwonnen worden, het stilzwijgende protest van de boerenbevolking nooit. Onderwijs dat niet aan hunne behoeften voldoet, zal eenvoudig niet bezocht worden, het zal doodbloeden."

21 *Handelingen der Staten-Generaal* [Proceedings of Parliament], October 2, 1919; in Dutch: "Het geheele onderwijs is doortrokken van de gedachte, dat de leerlingen aan die scholen naderhand boerinnen moeten worden en het ontleent daaraan zijn eigenaardig karakter. Niet alleen wordt aan die scholen in verschillende landbouw-

vakken onderwijs gegeven, doch ook het huishoudonderwijs is er geheel op gericht naderhand boerinnen te vormen. Men hoeft geen onderwijsspecialiteit te zijn om dit te begrijpen. Het ligt immers voor de hand, dat het huishoudonderwijs voor meisjes, die naderhand op de boerderij moeten komen, geheel anders is dan het huishoudonderwijs voor meisjes die later zullen komen, bv. in de stad in een middenstandsgezin."

22 This situation was caused by an amendment to include farm household management education on request of the association of urban household management teachers, without any warning. See also Burg, *"Geen tweede boer,"* 221–35.

23 For farm women's accounts, see, for instance, "Het leven van een Drentse Boerin [The life of a farm woman from the province of Drenthe]" (R.O K.), *In en om woning en boerderij* 35, no. 9 (September 1929): 140, 142; see also Margreet van der Burg and Krista Lievaart, *Drie generaties in schort en overall: Terugblik op een eeuw vrouwenarbied in de landbouw* [Three generations in aprons and overalls: An overview of a century of women's labour in agriculture] (Wageningen, Netherlands: Landbouwuniversiteit, 1998).

24 Openbare behandeling wetsontwerp staatsbegroting voor de landbouw, May 30, 1923; also quoted by Ament, *Verslagen van het Nederlandsch Landhuishoudkundig Congres* 76 (1924): 62; in Dutch: "Een gewone Brabantsche boer behoeft een levensgezellin, die thuis is in tuin—en hoenderstal—om die te verzorgen—meer dan een die tien verschillende puddingen in het hoofd heeft."

25 See Ine van Huet, "Tussen traditie en modernisering: Agrarisch bedrijf, boerinnentaak en landbouwhuishoudonderwijs in verandering, Winterswijk (Achterhoek) 1900–1940," [In between tradition and modernization: Agricultural business, farm women's tasks and farm household management education under change, Winterswijk (Achterhoek), 1900–1940] MA thesis, Catholic University Nijmegen 1988; Margreet van der Burg and Ine van Huet, "Niet het diploma, maar het huwelijk maakte je tot boerin": Onderwijs voor agrarische vrouwen vanaf 1850 tot 1940 ("Not the certificate but the marriage made me a farm woman": Education for farm women from 1850 until 1940), in *Een tien voor vlijt: Meisjesonderwijs vanaf de oudheid tot de MMS* [An A for Industriousness: Girls' education from ancient times to the advanced girls' school MMS], ed. K. den Dekker et al. (Walberg Pers, Netherlands: Zutphen, 1992), 117–34. The quotations are taken from the interviews held by Ine van Huet in 1987.

26 Huet, "Tussen traditie en modernisering," 49; in Dutch: "Wanneer zusters of nichten het hadden gevolgd, of vriendinnen, dan ging je daar ook heen. Je wilde daar bij horen."

27 See, for instance, Teacher from Gelderland, *Mededeelingen en Berichten der Gelders-Overijsselsche Maatschappij van Landbouw over 1921* (MB GOML 1921) [Announcements and messages of the Gelders-Overijssel Agricultural Society 1921], 4; Teacher from Zealand, *Verslag van de werkzaamheden* [Work report] 1920 (March 15, 1921), Archives of (Nieuw) Rollecate.

28 *25 jaar landbouwhuishoudonderwijs,* 66–68; in Dutch: "Het geleerde werd daar gedemonstreerd. . . . Succes hadden we! Veel belangstelling van de zijde der Drent-

sche vrouwen, die groote bewondering hadden voor al die prachtige vruchten in de glazen; 'hoe kan 't, hoe kan 't' hoorde men als een algemeene uitroep. Resultaat was: een der volgende jaren een flink bezette cursus."

29 Teacher from Gelderland, *MB GOML 1921*, 4, 6.

30 *In en Om de Woning en In en Om de Boerderij* 34 (1928), 176, in Dutch: "In onze afdeeling was geplaatst een lange werktafel, waaraan werd gekookt en waarop aan het eind de opgemaakte schotels waarvan niet werd geproefd, waren tentoongesteld, en verder een tafel, waaraan de bezoekers konden plaats nemen om de verschillende gerechten te proeven. Naar de bereiding van de schotels werd druk geïnformeerd en de kooksters lieten telkens opnieuw weer zien, hoe het geproefde werd klaargemaakt."

31 Teacher from Gelderland, *MB GOML 1920*.

32 Verslag van de werkzaamheden, March 1, 1920–February 20, 1921, Archives of (Nieuw) Rollecate. A regional newspaper, for instance, reported that twenty-eight Valburg pupils had a very successful excursion to the margarine factories in Oss. *De Betuwe* 25 (December 4, 1915).

33 Interview with a former teacher, 1988, quoted in Burg, *"Geen tweede boer,"* 247.

34 Interview with former pupil, Huet, "Tussen traditie en modernisering," 52, in Dutch: "Je had geen aanmoediging wat dat betreft, van mijn moeder wel, maar van mijn vader niet. Die wilde dat ouwe, grote bonen met spek of vlees en veel jus, flink vet allemaal en geen groenten."

35 Teachers meeting, October 29, 1934, Archives of (Nieuw) Rollecate.

36 *Vijf en twintig jaar Stichting voor Huishoudelijke Voorlichting ten Plattelande* [Twenty-five years of the Foundation for Rural Household Management Extension] (Staatsdrukkerij, Netherlands: Den Haag 1960), 19–20, in Dutch: "Doordat er dikwijls geen eigen slacht meer was, en er geen geld was voor bijkopen, werd de voeding steeds eenzijdiger. Men kon wel in aanmerking komen voor de z.g. goedkope levensmiddelen, zoals vlees en gehakt in blik, crisismargarine, enz. maar men voelde dit toch aan als het laatste, waarvan men gebruik zou maken."

37 Ibid., 16–17, in Dutch: "Dikwijls ondervond ik, dat zij niet wisten, hoe een ei te bakken, een eenvoudig beslag te maken, jus te maken, een smakelijke stamppot te bereiden, chocolademelk te koken, enz. Van kooktijden had men niet het minste idee. Bijv., rode kool móést 's morgens vroeg worden opgezet om om 12 uur 's middags gegeten te kunnen worden. Ook was het erg nuttig deze vrouwen iets te vertellen over voedingsleer, dit was voor haar een volkomen onbekend terrein en de belangstelling ervoor was meestal zeer groot."

38 Ibid., 17, in Dutch: "Wij gaven onze lessen in boerenschuren en—keukens, op zolders, in winkels, in afgekeurde schoollokalen, pakhuizen, consistoriekamers, raadzalen, burgemeesterkamers, café's."

39 *25 jaar landbouwhuishoudonderwijs*, 66, in Dutch:

> Eens bracht één van de vrouwen bessensap mee, die we op de les klaar zouden maken. We hadden tafels, stoelen, petroleumstellen en al, wat er verder

nodig was, op een grasveldje buiten gezet, omdat in het lokaal, waar anders de lessen werden gegeven, dien middag een vergadering werd gehouden. Al vertellende probeerde ik de flesch te ontkurken. Nog vóór de kurk er echter uitgetrokken was, sprong deze er met een knal uit en spoot het kostelijke nat met stralen over de aandachtig luisterende gezichten van mijn leerlingen. Van rood tot blauw en pimpelpaars waren in een ommezien wangen en haren geverfd.

40 *25 jaar Stichting voor Huishoudelijke Voorlichting ten Plattelande*, 26–27.

41 Ibid., 28; in Dutch: "De lerares N XIX demonstreerde dan iets op het gebied van koken en ik, Na-lerares, bekeek de oude kleren welke meegebracht waren."

42 Manuscript, radio lecture, VARA, January 1, 1940, 8, Archives of (Nieuw) Rollecate; foods that could be gathered in the wild mentioned in Dutch: zeekraal, lamsoor, bramen, vlierbessen en bosbessen.

43 *25 jaar landbouwhuishoudonderwijs*, 41–42.

44 *25 jaar Stichting voor Huishoudelijke Voorlichting ten Plattelande*, 16–17; in Dutch: "Kwamen er steeds vragen los op allerlei gebied: op wasgebied, hoe kledingstukken te veranderen of te verstellen, hoe kapot speelgoed te repareren, hoe eenvoudige versnaperingen te maken, enz."

45 Farm household management teacher, report c. 1915, Rijksarchief (RA) Groningen, Groninger Maatschappij van Landbouw, 1878–1940 [Groninger Agricultural Society], inv. 947, 4, 25.

46 Ibid.

47 Report of visit to former pupils' courses at Usselo en Boerhaar, 1920, Archives of (Nieuw) Rollecate; in Dutch: "Al veel verbeterd is, maar dat er ook nog veel is, wat we gaarne anders zouden wenschen. Misschien zal dit ook mettertijd nog wel veranderen, vooral nu de boerin blijk heeft gegeven, dat zij wel eens iets veranderen wil en dat zij bekennen wil, dat zij zich bij die verandering goed bevindt."

48 Farm household management teacher, posted at Zwolle, Archives of (Nieuw) Rollecate.

49 Farm household management teacher, posted at Hengelo, report of practical results of farm household management education, household part, 1920, Archives of (Nieuw) Rollecate.

50 Farm household management teacher, posted at Zwolle, Archives of (Nieuw) Rollecate; in Dutch: "Daar waar de vrouwelijke plattelandsbevolking nog zoo'n groot deel harer werkkrachten aan het bedrijf moet geven, is het niet mogelijk meer tijd te besteden aan de bereiding der maaltijden enz. . . . En uit de aard der zaak zal ook de inrichting der woning eenvoudig moeten zijn, om zo weinig mogelijk onderhoud te vragen. Toch is het een verblijdend teeken dat men een andere inrichting wenscht, als de eenvoudige woonkeukens met open vuren."

51 Ibid.

52 Huet, "Tussen traditie en modernisering," 57.

53 See, for instance, Burg and Huet, "Niet het diploma."

54 Farm household management teacher, posted at Zwolle, Archives of (Nieuw) Rollecate.

55 Teacher from Gelderland, *MB GOML 1919*, 68–69; Teacher from Gelderland, *MB GOML 1920*, 6.

56 Farm household management teacher, Report about Posterholt in 1920, April 20, 1921, Archives of (Nieuw) Rollecate, in Dutch: "Soms is de leerstof moeilijk voor hen, daar ze allen zoo vroeg mogelijk van school zijn gekomen en daarna thuis steeds hard hebben moeten meewerken, in de boerderij, zowel thuis als op het land. Zodoende hebben ze weinig kennis van de huishoudelijke vakken. De meesten hebben zelden of nooit gekookt, wat door het volgen van de cursus een beetje verandert in die huishoudingen, waar de vrouw des huizes het roer wel een klein beetje uit handen wil geven."

57 See, for instance, Heide Inhetveen, "Women Pioneers in Farming: A Gendered History of Agricultural Progress," *Sociologia Ruralis* 38 (1998): 265–85; for women in agricultural sciences in the Netherlands, see Margreet van der Burg and Marian Bos-Boers, *Vrouwen, Wageningen en de Wereld: Wetenschap, studie en loopbaan, 1918–2003* [Women, Wageningen and the world: Science, study and career, at Wageningen Agricultural College/University1918–2003] (Hilversum, Netherlands: Verloren, 2003).

Chapter 4

1 Harriette E. Cushman, "The Forgotten Hen," Montana State University Archives, Bozeman, Montana, Montana Extension Service, accession 00021, box 87, folder 47, *Report of the State Extension Poultry Specialist*, 1935, 51; hereafter cited as *Annual Poultry Report*.

2 *Annual Poultry Report*, 1922, 5.

3 Letter to Jackson Martindell, Special Collections, Montana State University, Bozeman, Montana. Merrill G. Burlingame Papers, 1880–1900, Collection 2245, box 41, folder 17.

4 *Annual Poultry Report*, 1922, 5.

5 *Annual Poultry Report*, 1923, 6.

6 "Twenty-Five Year History of the Poultry Extension Program of Montana," *Annual Poultry Report*, 1939, 25.

7 Ibid.

8 Ibid., 26.

9 *Annual Poultry Report*, 1921, 2.

10 *Annual Poultry Report*, 1922, 10.

11 *Annual Poultry Report*, 1923, 10.

12 *Annual Poultry Report*, 1924, 11.

13 Candling is "the process of viewing the interior of an egg before a bright source of light in a darkened room." "Poultry Production Goals in the Montana State Defense Program," *Annual Poultry Report*, 1942, 3.

14 James T. Sparling, Commissioner of Agriculture, and B. F. Thrailkill, Chief, Dairy Division, "Eggs: Montana State Egg Law, Grades, Rules and Regulations," Montana Department of Agriculture, Dairy Division, 1937, 19.

15 "The 'Montana Extra Select,'" Poultry Hints 4, no. 3 (March 1924), in *Annual Poultry Report*, 1924, 42.

16 Harriette E. Cushman, "'Montana Extra Selects,' or Montana Eggs for Montana People," *Montana Woman*, November 1924, 22.

17 *Annual Poultry Report*, 1924, 11.

18 "County Sends Egg Exhibits to Butte Show," newspaper clipping, *Annual Poultry Report*, 1924, 51.

19 "Exhibits for Egg Show Are Placed," *Annual Poultry Report*, 1924, 51; "Eggs and Chicks of Poultry Show Crowding Chamber," *Annual Poultry Report*, 1924, 53.

20 "County Sends Egg Exhibits to Butte Show," newspaper clipping, *Annual Poultry Report*, 1924, 51.

21 *Annual Poultry Report*, 1924, 54.

22 Ibid., 59.

23 Ibid., 60.

24 *Annual Poultry Report*, 1926, 13.

25 *Annual Poultry Report*, 1924, 42.

26 *Annual Poultry Report*, 1925, 14.

27 *Annual Poultry Report*, 1924, 12; *Montana Woman*, November 1924, 22.

28 *Montana Woman*, November 1924, 22.

29 Montana State University Archives, Bozeman, Montana, Agricultural Economics Department, accession 00002, box 23, folder 10; hereafter cited as Agricultural Economics Papers. Letter to M. L. Wilson, March 22, 1924. In 1933 Wilson went on to serve as the chief economist in the Agricultural Adjustment Administration.

30 Agricultural Economics Papers, box 23, folder 10, "Minimum Equipment."

31 Agricultural Economics Papers, box 23, folder 10 "Regulations Concerning the Grading, Marking and Packing of Montana Eggs."

32 "Twenty-Five Year History," *Annual Poultry Report*, 1939, 30.

33 *Annual Poultry Report*, 1931, 50.

34 "Twenty-Five Year History," *Annual Poultry Report*, 1939, 31.

35 "Twenty-Five Year History," *Annual Poultry Report*, 1938, 31.

36 *Annual Poultry Report*, 1940, 79.

37 "Montana's Egg Quality Demonstration," *Annual Poultry Report*, 1940, 1.

38 Ibid.

39 *Annual Poultry Report*, 1944, 59.

40 Ibid.

41 *Annual Poultry Report*, 1953, 55.

42 Harriette E. Cushman, "Know Your Eggs," Montana Extension Service in Agriculture and Home Economics Folder, no. 14 (June 1951), 2.

43 *Annual Poultry Report*, 1952, 75.

44 *Annual Poultry Report*, 1932, 49.

45 *Annual Poultry Report*, 1942, 65.

46 Ibid., 66.

Chapter 5

1 University of Massachusetts Amherst, Special Collections and University Archives, RG 2/2, MAC Board of Trustees, Minutes, October 2, 1912.

2 Margaret Rossiter, "Women Scientists in America before 1920," *American Scientist* 62, no. 3 (May–June 1973): 316.

3 Julie N. Zimmerman and Olaf F. Larson, *Opening Windows onto Hidden Lives: Woman, Country Life, and Early Rural Sociological Research* (Univeristy Park: Pennsylvania State University Press, 2010), 10.

4 Lowry Nelson, *Rural Sociology: Its Origin and Growth in the United States* (Minneapolis: University of Minnesota Press, 1969), 11.

5 Gene Wunderlich, *American Country Life* (Dallas: University Press of America, 2003), 29.

6 MAC, Annual Report, 1908, 24.

7 Jefferson taught this class within the Humanities Division. In the 1914 annual report, the head of the division noted that Miss Jefferson was arguing for the class to be two semesters, but he did not agree. The airing of this disagreement in print reflects Jefferson's connection to Butterfield and her influence with Butterfield.

8 John R. Commons et al., *History of Labour in the United States* (1918–35; repr. New York: Augustus M. Kelley, 1966), x.

9 Alexander E. Cance and Lorian P. Jefferson, "Farmer's Cooperative Corporations," *Bulletin of the Vermont Department of Agriculture*, no. 24 (December 1915).

10 Lorian P. Jefferson, "The California Fruit Growers' Exchange," *Business America* 13, no. 4 (April 1913): 330.

11 Cance and Jefferson, 5.

12 Lorian Jefferson, "The Costs of Marketing the Apple Crop of 1923," Massachusetts Agricultural Experiment Station, Bulletin no. 224 (March 1925).

13 Lorian Jefferson, "The Market Outlet for Massachusetts Apples," Massachusetts Agricultural Experiment Station, Bulletin no. 231 (January 1927).

14 Ibid., 39.

15 Lorian Jefferson, "The McIntosh Apple on the New York Market," Massachusetts Agricultural Experiment Station, Bulletin no. 243 (May 1928).

16 Ibid., 200.

17 Frederick V. Waugh, "Quality Factors Influencing Vegetable Prices," *Journal of Farm Economics* 10 (1928): 187.

18 University of Massachusetts Amherst, Special Collections and University Archives, FS 72, Lorian Jefferson Papers, Lorian Jefferson, "The Future of the Connecticut Valley Onion Industry," unpublished copy.

19 Lorian Jefferson, "Farm Ownership in Massachusetts," *Journal of Farm Economics* 4 (October 1923) / Massachusetts Agricultural Experiment Station, Contribution no. 18 (1923).

20 Lorian P. Jefferson, "The Balance of Trade in Farm Products," *Journal of Farm Economics* 8 (1926) / MAC Experiment Station Bulletin No. 51 (1926).

21 Ibid.

22 Jefferson, "Balance of Trade," 454.

23 Lorian P. Jefferson, "The Community Market," MAC Extension Bulletin no. 21, Amherst (April 1918).

24 Alexander Cance Papers, FS 45, box 5, folder 50.

25 Jefferson, "Farm Ownership in Massachusetts," MAC Experiment Station (1923): 214.

26 Glenda Riley, *Inventing the American Woman: An Inclusive History since 1877* (Wheeling, IL: Harlan Davidson, 2001), 315.

27 Cance, "The Decline of the Rural Population in New England," *Publications of the American Statistical Association* 13, no. 97 (March 1912): 100.

28 Barbara Miller Solomon, *In the Company of Educated Women: A History of Women and Higher Education in America* (New Haven, CT: Yale University Press, 1985), 63.

29 MAC Annual Report, 1914, 31.

30 University of Massachusetts Amherst, Special Collections and University Archives, Margaret P. Hamlin Faculty file, RG 40/11, Margaret Hamlin, "Pioneering in Placement for Women in Agriculture." Hamlin noted the post-WWII focus shifted from agriculture and horticulture to other academic areas. She resigned in 1948.

31 Schlesinger Library, Radcliffe Institute, Harvard University, Cambridge, MA, Woman's National Farm and Garden Association Additional Records, 1913–2008; MC 594, ser. 4 (17.1).

32 Martha A. Nolan, *A Chronicle: The History of Woman's National Farm and Garden Association, Incorporated, 1914–1984* (Fremont, OH: Lesher Printers, 1985), 6.

33 Boston University Archives, BU13, Charlotte Barrell Ware Papers, box 11, folder B11.

34 Harvard University Library Open Collections Program, Women Working 1800–1930, "Women's Educational and Industrial Union," http://ocp.hul.harvard.edu/ww/weiu.html. Founded in 1911–12 in New York City as alumnae associations of what later became the Seven Sisters' colleges, the Intercollegiate Bureau of Occupations focused on securing employment for college women and using the organization's power to develop opportunities for women.

35 Elaine F. Weiss, *Fruits of Victory: The Women's Land Army of America in the Great War* (Washington, DC: Potomac Books, 2008), 33.

36 *Massachusetts Daily Collegian*, October 16, 1918.

37 University of Massachusetts Amherst, Special Collections and University Archives, Kenyon L. Butterfield Papers, RG 3/1, box 20, folder 300. On a list of prospective members of the Advisory Council for Women, Butterfield listed "Miss Jefferson. Well acquainted with leaders in Farm and Garden Association."

38 University of Massachusetts Amherst, Special Collections and University Archives, Student Affairs Records, RG 30/2, Nettie Caroline Burleigh, "Dedication of the Dormitory," *Women in Agriculture and Country Life*, October 7–8, 1920. According to her alumni records, Burleigh's father had one of the first established herds of Hereford cattle in the United States on his farm, also located in Waterville. Burleigh would write an account of her decision to become a farmer and her farming activities in a special edition of the WNFGA *Bulletin* published in 1924, edited by Jefferson and designated as a "Special New England Number."

39 Mount Holyoke College Archives, Nettie C. Burleigh Records.

40 University of Massachusetts Amherst, Special Collections and University Archives, Lorian P. Jefferson Papers, FS 072, Nettie Caroline Burleigh, "Some Experiences of a Woman Farmer," *Farm and Garden: Special New England Number* 11 (April 1924), 22.

Chapter 6

I presented a paper at the Agricultural History Society meeting in Banff, Alberta, in 2013 that formed the basis of part of this chapter; it was later revised for publication as "The New Woman in Rural British Columbia: Madge Robertson Watt and the Women's Institutes, 1893–1913," *Pacific Northwest Quarterly* 105 (Winter 2013/14): 3–11.

1 Photographs of the infamous cake-cutting incident were widely circulated in the British press. See, for example, *The Daily Mail*, http://www.dailymail.co.uk/femail/article-3110804/Still-spritely-despite-declared-dead-BBC-Queen-smiles-Sophie-Anne-meet-members-W-I.html. Readers will recognize a parallel to the Canadian and British Women's Institutes in groups created for rural women by the U.S. Department of Agriculture through state extension services such as those in Arkansas. See Cherisse Jones-Branch, chap. 8 in this volume. Another parallel is the community-based, state-sponsored instruction extended to groups of rural women in the Netherlands. See Margreet van der Burg, chap. 3 in this volume.

2 Maggie Andrews, *The Acceptable Face of Feminism: The Women's Institute as a Social Movement* (London: Lawrence and Wishart, 1997; rev. 2nd ed., 2015); and Joan M. Jensen, *Calling This Place Home: Women on the Wisconsin Frontier, 1850–1925* (St. Paul, MN: Minnesota Historical Society, 2006), 127–30.

3 Nancy Tomes, "Spreading the Germ Theory: Sanitary Sciences and Home Economics, 1880–1930," in *Rethinking Home Economics: Women and the History of a Profession*, ed. Sarah Stage and Virginia B. Vincenti (Ithaca: Cornell University Press, 1997), 35.

4 *Cake Bakers and Trouble Makers: Lucy Worsley's 100 Years of the WI* (London: BBC Two, 2015).

5 Adelaide Hoodless, *Public School Domestic Science* (Toronto: Copp Clark, 1898), 31.

6 Ibid., 30. The book includes sections on "Suggestions for Young Housekeepers," "Caring for Invalids," and "Infants' Diet."

7 See, for example, Margaret Kechnie, "Defining the Lives of Rural Women: Laura Rose on 'The Womanly Sphere of Woman,'" in *Framing Our Past: Canadian Women's History in the Twentieth Century*, ed. Sharon Anne Cook, Lorna R. McLean, and Kate O'Rourke (Montreal: McGill-Queen's University Press, 2001), 113.

8 Tomes, "Spreading the Germ Theory," 34–54; and Nancy Tomes, *The Gospel of Germs: Men, Women, and the Microbe in American Life* (Cambridge, MA: Harvard University Press, 1998).

9 Andrews takes up this question about domesticity and feminism among contemporary young women in the United Kingdom today, in "Afterword: Domesticity,

Feminism, and the WI in the Twenty-First Century," in *The Acceptable Face of Feminism*, 223–42.

10 For more on the Ontario Women's Institute, see Linda M. Ambrose, *For Home and Country: The Centennial History of the Women's Institutes in Ontario, 1897–1997* (Erin, ON, 1996); and Margaret Kechnie, *Organizing Rural Women: The Federated Women's Institutes of Ontario, 1897–1919* (Montreal and Kingston: McGill-Queen's University Press, 2003).

11 On the British Columbia WI, see Jill Douglas, ed., *Modern Pioneers: 1900–1959* (Victoria, BC: British Columbia Women's Institutes, 1959); Carol J. Dennison, "They Also Served: The British Columbia Women's Institutes in Two World Wars," in *Not Just Pin Money: Selected Essays on the History of Women's Work in British Columbia*, ed. Barbara K. Latham and Roberta J. Pazdro (Victoria, BC: Camosun College, 1984), 211–19; Alexandra Zacharias, "British Columbia Women's Institutes in the Early Years: Time to Remember," in *In Her Own Right: Selected Essays on Women's History in B.C.*, ed. Barbara Latham and Cathy Kess (Victoria, BC: Camosun College, 1980), 55–78; and Linda M. Ambrose, *Women's Institutes in Canada: The First One Hundred Years, 1897–1997* (Ottawa, ON: Tri-County, 2000), 100–103. For more on American groups that parallel the WI, see, for example, Dorothy Schwieder, *75 Years of Service: Cooperative Extension in Iowa* (Ames: Iowa State University Press, 1993); Katherine Jellison, *Entitled to Power: Farm Women and Technology, 1913–1963* (Chapel Hill, NC: University of North Carolina Press, 1993), 16–65; and Jenny Barker-Devine, "Rural Women's Activism," American Agricultural History Primer, Iowa State University Center for Agricultural History and Rural Studies, http://rickwoten.com/RuralWomensActivism.html.

12 Linda M. Ambrose, "'What Are the Good of Those Meetings Anyway?' Early Popularity of the Ontario Women's Institutes," *Ontario History* 87, no. 1 (Spring 1995), 1–19.

13 Jo Burr Margadant, "Introduction: Constructing Selves in Historical Perspective," in *The New Biography: Performing Femininity in Nineteenth-Century France*, ed. Jo Burr Margadant (Berkeley: University of California Press, 2000), 7.

14 Judith Butler, "Performative Acts and Gender Constitution: An Essay in Phenomenology and Feminist Theory," in *Performing Feminisms: Feminist Critical Theory and Theatre*, ed. Sue-Ellen Case (Baltimore, MD: Johns Hopkins University Press, 1990), 270–82.

15 Royal Commission on Industrial Training and Technical Education, *Report of the Commissioners*, 4 vols. (Ottawa, 1913) 4:2352.

16 Linda M. Ambrose and Kristin Hall, "A New Woman in Print and Practice: The Canadian Literary Career of Madge Robertson Watt, 1890–1907," *History of Intellectual Culture* 7 (2007), http://www.ucalgary.ca/hic/issues/vol7.

17 On transportation in, and the historical geography of, British Columbia, see Cole Harris, "Moving amid the Mountains, 1870–1930," *B.C. Studies* 58 (Summer 1983), 3–39. On rural British Columbia social history, see R. W. Sandwell, ed., *Beyond the City Limits: Rural History in British Columbia* (Vancouver: University of British Columbia Press, 1999); and Richard Somerset Mackie, *Island Timber: A Social History of the Comox Logging Company, Vancouver Island* (Victoria: Sono Nis Press, 2000).

18 Department of the Secretary of State, *The Civil Service List of Canada, 1897* (Ottawa, 1897), 144, available online at Canadiana, University of Alberta Archives, http://ebooks.library.ualberta.ca/local/cihm_46489. To convert the salary, I used the conversion tool at Samuel H. Williamson, "Seven Ways to Compute the Relative Value of a U.S. Dollar Amount, 1774 to Present," MeasuringWorth, http://www.measuringworth.com/uscompare/. My thanks to Andrew Smith for suggesting these conversion tools.

19 Library and Archives Canada, Fourth Census of Canada, 1901 (Ottawa, 1902), province British Columbia, district Victoria, subdistrict Metchosin, Dwelling house 17, p. 2, item no. 158233; available online at http://data2.collectionscanada.ca/1901/z/z001/pdf/z000011935.pdf; Library and Archives Canada, Fifth Census of Canada, 1911, 6 vols. (Ottawa, 1912–15), province British Columbia, district Victoria, subdistrict Metchosin, Dwelling house 26, p. 3, item no. 678819, available online at http://data2.collectionscanada.gc.ca/1911/pdf/e001938087.pdf. For more demographic context about how the Watts' income level and household composition compared to other British Columbians, see John Douglas Belshaw, *Becoming British Columbia: A Population History* (Vancouver, BC: University of British Columbia Press, 2009).

20 David Demeritt, "Visions of Agriculture in British Columbia," *BC Studies* 108 (Winter 1995–96): 36, 37.

21 *Victoria Times*, May 5, 1905, 6, and April 15, 1910, 7, 12.

22 *Victoria Times*, September 27, 1910, 5.

23 Kechnie, "Defining the Lives of Rural Women," 113.

24 Laura Rose, *Farm Dairying* (Chicago: A. C. McClurg, 1911). One indication of the book's wide use is that it had been published in four editions by 1916.

25 Kechnie, "Defining the Lives of Rural Women," 113.

26 Ambrose, *Women's Institutes in Canada*, 100–101.

27 *Victoria Times*, October 8, 1909, 3.

28 For more on the Ontario WI, see Ambrose, *For Home and Country*; and Kechnie, *Organizing Rural Women*.

29 Ambrose, *For Home and Country*, 32.

30 Canada, Royal Commission on Industrial Training and Technical Education, *Report of the Commissioners*, 4 vols. (Ottawa, 1913), 4:2353.

31 For more on the Agricultural Instruction Act, see Linda M. Ambrose, "Better and Happier Men and Women: The Agricultural Instruction Act, 1913–1924," *Historical Studies in Education / Revue d'histoire de l'éducation* 16 (Fall 2004): 257–85.

32 BC Archives, Advisory Board of Women's Institutes of British Columbia, Women's Institute, microfilm reel A1660, Minutes and Papers, 1911–32, "Notes on Women's Institutes in B.C.," 1.

33 *Victoria Times*, December 6, 1910, 4.

34 Ibid. See also *Report of the Commissioners* 4:2349. The speaker, Annie Cleland, estimated that between 3,000 and 4,000 women in Victoria (out of the total population of 50,000) were working for pay. Women's contributions to nation building in British Columbia is a theme of Jean Barman, *Sojourning Sisters: The Lives of Jessie and Annie McQueen* (Toronto: University of Toronto Press, 2003).

35 *Report of the Commissioners* 4:2352–53.

36 Ibid., 4:2352.

37 Ibid.

38 Ambrose, *Women's Institutes in Canada,* 100–103; Linda M. Ambrose, *A Great Rural Sisterhood: Madge Robertson Watt and the ACWW* (Toronto: University of Toronto Press, 2015), 69–96.

39 BC Archives, Advisory Board of Women's Institutes of British Columbia, Women's Institute, microfilm reel A1660, Minutes and Papers, 1911–32, "Minutes of Meeting of Advisory Board of Women's Institutes," August 14 and 15, 1911, 1.

40 Ibid., 2.

41 Ibid.

42 Ibid.

43 Kechnie, *Organizing Rural Women,* 12.

44 Ibid.

45 "Minutes of Meeting of Advisory Board of Women's Institutes," August 14 and 15, 1911, 9.

46 For examples of the column see *Toronto Globe,* January 21, 1893, 6; January 28, 1893, 6; February 11, 1893, 6; February 18, 1893, 6; and February 25, 1893, 5.

47 Peter A. Baskerville, "'She Has Already Hinted at "Board"': Enterprising Urban Women in British Columbia, 1863–1896," *Histoire sociale / Social History* 26 (November 1993): 205–27; Peter A. Baskerville, "Women and Investment in Late-Nineteenth-Century Urban Canada: Victoria and Hamilton, 1880–1901," *Canadian Historical Review* 80 (1999), 191–218; Peter A. Baskerville, *A Silent Revolution? Gender and Wealth in English Canada, 1860–1930* (Montreal: McGill-Queen's University Press, 2008).

48 On feminism in the Canadian West, see Kathryn McPherson, "Was the 'Frontier' Good for Women? Historical Approaches to Women and Agricultural Settlement in the Prairie West, 1870–1925," *Atlantis* 25 (Fall/Winter 2000): 75–86; and Veronica Strong-Boag, "Pulling in Double Harness or Hauling a Double Load: Women, Work and Feminism on the Canadian Prairie," *Journal of Canadian Studies* 21 (Fall 1986): 32–52.

49 "Minutes of Meeting of Advisory Board of Women's Institutes," August 14 and 15, 1911, 1.

50 "Minutes," April 29, 1910, p. 9.

51 Ibid.

52 *Victoria Times,* April 11, 1911, 8.

53 Again, this is a departure from the Ontario WI model, where any discussion of political affairs was discouraged. See Ambrose, "What Are the Good of Those Meetings Anyway?" 1–19.

54 *Victoria Times,* August 10, 1912, 5, and August 24, 1912, 10.

55 Ambrose, *A Great Rural Sisterhood,* 110.

56 Ibid., 6, 17, 98, 123, 126, 234–35.

57 Andrews, *The Acceptable Face of Feminism,* 43–52; Ambrose, *A Great Rural Sisterhood,* 97–123.

58 See Ambrose, *A Great Rural Sisterhood*, 116–23.

59 Andrews, *The Acceptable Face of Feminism*, 115; and Maggie Andrews, "The WI's Rural Retailing and Markets 1915–1939: A First World War Legacy," *History of Retailing and Consumption* 1, no. 2 (2015): 89–104. See also, "Country Markets," http://www.country-markets.co.uk/about-us.

60 See Ambrose, *A Great Rural Sisterhood*, 124–69.

61 Associated Country Women of the World, http://www.acww.org.uk/about_news/article/108.

62 Associated Country Women of the World, http://www.acww.org.uk/agriculture_news_events/article/25.

63 Associated Country Women of the World, http://www.acww.org.uk/agriculture_news_events/article/26.

Chapter 7

1 *Radio Times*, April 2, 1935.

2 Judith Butler, *Gender Trouble* (London: Routledge, 1990).

3 For further discussion on centrality of the rural to notions of the rural and Englishness, see Alun Howkins, "The Discovery of Rural England," in *Englishness: Politics and Culture 1880–1920*, ed. P. Dodd and R. Colls (London: Croom Helm, 1987), 78–92.

4 M. E. Webb, *Farmhouse Cookery* (London: George Newnes, 1936), 47.

5 S. Moores, *Media and Everyday Life in Modern Society* (Edinburgh: Edinburgh University Press, 2000).

6 *Radio Times*, November 18, 1938, 17.

7 *Radio Times*, November 16, 1934, 546.

8 Interview with Mrs. Lynch, Portsmouth, May 9, 2002.

9 For further discussion of radio and domesticity, see Maggie Andrews, *Domesticating the Airwaves: Broadcasting, Domesticity and Femininity* (London: Continuum, 2012).

10 *Radio Times*, April 12, 1937, 10.

11 Alun Howkins, *The Death of Rural England: A Social History of the Countryside since 1900* (London: Routledge, 2003), 82.

12 *The Times* (London), broadcasting supp., August 14, 1934, iv.

13 *Radio Times*, March 24, 1939. Quoted in Maggie Andrews, *Domesticating the Airwaves: Broadcasting, Domesticity and Femininity* (London: Continuum, 2012), 5.

14 *BBC Hand Book 1928*, 125.

15 Moores, *Media and Everyday Life*, 47.

16 C. Frederick, *Scientific Management in the Home: Household Engineering* (Chicago: American School of Home Economics, 1920).

17 K. Lacey, *Feminine Frequencies: Gender, German Radio, and the Public Sphere 1923–1945* (Ann Arbor: University of Michigan Press, 1996), 154.

18 K. Laybourn, *Modern Britain since 1906: A Reader* (London: Tauris, 1999), 105.

19 E. Roberts, *A Woman's Place: An Oral History of Working Class Women 1890–1940* (Oxford: Blackwell, 1984), 161.

20 For further discussion on the increasing influence of a nutritional discourse around food, see A. Beardsworth and T. Keil, *Sociology of the Menu* (London: Routledge, 1997).

21 British Broadcasting Company Written Archives at Caversham, England (hereafter BBCWA), Mottram file, BBC to Mottram, January 27, 1931.

22 BBCWA, Mottram to the BBC, July 23, 1935.

23 Michael Bailey, *Narrating Media History* (London: Routledge, 2009), 63.

24 *Radio Times*, December 1, 1937.

25 Bertrand Russell, *Autobiography* (1967–1969); rpt. Oxford: Routledge, 2009.

26 *Radio Times*, December 17, 1935.

27 J. M. Winter, *The Great War and British People*, 2nd ed. (Basingstoke: Palgrave Macmillan, 2003), 215.

28 Sylvia Pankhurst described the hardship faced by women in the East End as they struggled with price rises and shortages in the first years of the war; Pankhurst, *The Home Front* (1932; London: Cresset, 1987).

29 G. Robb, *British Culture and the First World War* (Basingstoke: Palgrave MacMillan, 2002).

30 S. Hockenhull, "Everybody's Business: Film, Food and Victory in the First World War," *Historical Journal of Film, Radio and Television* 35, no. 4 (2013), 11. For further discussion of the discontent and politicization around food supplies, see K. Hunt, "The Politics of Food and Women's Neighbourhood Activism in First World War Britain," *International Labour and Working-Class History* 77 (2010): 8–26.

31 For a more detailed discussion, see L. M. Barnett, *British Food Policy during the First World War* (London: Routledge, 1985).

32 T. George, "Female Agricultural Workers in Wales in the First World War," in *The Home Front in Britain: Images, Myths and Forgotten Experiences since 1914*, ed. M. Andrews and J. Lomas (Basingstoke: Palgrave Macmillan, 2014), 92–107.

33 N. Mansfield, *English Farmworkers and Local Patriotism, 1900–1930* (Aldershot: Ashgate, 2001).

34 M. Andrews, A. Gregson, and J. Peters, *Voices of the First World War: Worcestershire's War* (Stroud: Amberley, 2014).

35 P. E. Dewey, *British Agriculture in the First World War* (1989; London: Routledge, 2014), 53.

36 Ibid., 54.

37 *Coventry Evening Standard*, August 11, 1917.

38 For further discussion of the formation of the Women's Institute movement in Britain, see M. Andrews, *The Acceptable Face of Feminism* (1997; London: Lawrence and Wishart, 2015).

39 *Whitby Gazette*, March 9, 1917.

40 Worcestershire Archives and Archaeology Service, 5584, ref. 705:677, parcel 1.

41 For more details on horticultural colleges for women, see A. Meredith, "Horti-

cultural Education in England, 1900–40: Middle-Class Women and Private Gardening Schools," *Garden History* 31, no. 1 (Spring 2003): 67–79.

42 A. Davin, "Imperialism and Motherhood," *History Workshop Journal* 5, no. 1 (1978): 9–66.

43 Roberts, *A Woman's Place*, 151.

44 Judy Giles, *The Parlour and the Suburb: Domestic Identities, Class, Femininity and Modernity* (London: Berg, 2004), 117.

45 *Radio Times*, September 6, 1932.

46 For discussion of the working-class standard of living in Britain, see I. Gazeley and A. Newell, "The First World War and Working-Class Food Consumption in Britain," *European Review of Economic History* 17 (2013): 71–94; and I. Gazeley and S. Horrell, "Nutrition in the English Agricultural Labourer's Household over the Course of the Long Nineteenth Century," *Economic History Review* 66, no. 3 (2013): 757–84.

47 M. E. Webb, *Economical Cooking* (London: Philip, 1934), 71.

48 The *Farmers Weekly* photograph collection is available at https://www.reading.ac.uk/merl/collections/Archives_A_to_Z/merl-P_FW.aspx. For a discussion of *Farmers Weekly*'s approach to domestic womanhood, see N. Verdon, "The Modern Countrywoman, Domesticity and Social Change in Interwar Britain," *History Workshop Journal* 70 (2010): 78–107.

49 A. King, "From Sage on the Stage to Guide on the Side," *College Teaching* 41, no. 1 (1993): 30–35.

50 Parliamentary Recruiting Committee 1915, University Library Special Collections Leeds, http://library.leeds.ac.uk/special-collections-exhibitions-war-propaganda#activate-image7.

51 Howkins, "The Discovery of Rural England."

52 J. Bourke, *Working Class Cultures in Britain 1890–1960* (Oxford: Routledge, 1996), 168.

53 Ibid.

54 Ibid., 163.

55 R. J. Moore-Colyer, "From Great Wen to Toad Hall: Aspects of the Urban-Rural Divide in Inter-War Britain," *Rural History* 10, no. 1 (1999): 105–24, 107.

56 Howkins, *The Death of Rural England*.

57 Moore Colyer, "From Great Wen to Toad Hall," 112.

58 For a more extension discussion, see Diane Tye, *Baking as Biography: A Life Story in Recipes* (Montreal: McGill-Queen's University Press, 2010).

59 BBC *Household Talks*, February 28, 1933.

60 D. Cardiff, "The Serious and the Popular: Aspects of the Evolution of Style in Radio Talk 1928–1939," in *Media Culture and Identity: A Critical Reader*, ed. R. Collins (London: Sage, 1986), 228–41, 237.

61 *Household Talks*, April 21, 1934.

62 John Boyd Orr, *Food Health and Income: Report on a Survey of Adequacy of Diet in Relation to Income* (London: Macmillan, 1937).

63 *Radio Times*, October 30, 1934.

64 *Household Talks*, "How I Manage," September 11, 1934.

65 M. Tebbutt, *Women's Talk? A Social History of 'Gossip' in Working Class Neighbourhoods, 1880–1960* (Aldershot: Scholars' Press, 1995), 11.

66 *Household Talks*, April 2, 1935.

67 Ibid.

68 Ibid.

69 H. Jennings and W. Gill, *Broadcasting in Everyday Life: A Survey of the Social Effects of the Coming of Broadcasting* (London: BBC Publications, 1939), 12.

70 Webb, *Farmhouse Cookery*, 60.

71 Ibid., 71.

72 Ibid., 60.

73 P. Thane and J. Mackay, "The Englishwoman," in *Englishness, Politics and Culture*, ed. P. Dodd and R. Colls (London: Croom Helm, 1986).

74 Webb, *Farmhouse Cookery*, 47.

75 Jennings and Gill, *Broadcasting in Everyday Life*, 74.

76 A. Crisell, *An Introductory History of British Broadcasting* (London: Routledge, 1997), 17.

77 BBCWA Mrs. Webb file.

Chapter 8

Parts of this chapter draw on my essay "Empowering Families and Communities: African American Home Demonstration Agents in Arkansas, 1913–1965," in *Race and Ethnicity in Arkansas: New Perspectives*, ed. John A. Kirk, 85–95 (Fayetteville: University of Arkansas Press, 2014). The University of Arkansas Press generously granted its permission to reprint those sections in this collection.

1 Mary S. Hoffschwelle, "'Better Homes on Better Farms': Domestic Reform in Rural Tennessee," *Frontiers: A Journal of Women Studies* 22, no. 1 (2001): 52.

2 Stephen F. Strausberg, *A Century of Research: A Centennial History of the Arkansas Agricultural Experiment Station* (Fayetteville: Arkansas Agricultural Experiment Station, University of Arkansas, 1989), 30; Alfred Charles True, *A History of Agricultural Education in the United States, 1785–1925* (New York: Arno Press, 1969), 280–81, 365.

3 Jeannie M. Whayne, Thomas A. DeBlack, George Sabo II, and Morris S. Arnold, eds., *Arkansas: A Narrative History* (Fayetteville: University of Arkansas Press, 2002), 308.

4 Florence E. Ward, "Home Demonstration Work under the Smith-Lever Act, 1914–1924," Circular 43 (Washington, DC: U.S. Department of Agriculture, June 1929); Hoffschwelle, 53.

5 Frank Lincoln Mather, ed., *Who's Who of the Colored Race: A General Biographical Dictionary of Men and Women of African Descent* (Memento Edition, Half-Century Anniversary of Negro Freedom in U.S., Chicago, 1915), 1:227. Ray also studied vocational education and dairying at Kansas Agricultural College in Manhattan and was the

director of agriculture at Langston University in Oklahoma until he moved to Little Rock, Arkansas. See the *Kansas State Agricultural College Catalogue, Fifty-First Session, 1913–1914* (Topeka: Kansas State Printing Office, 1914), 354. According to one source, there was at least one black agent in Arkansas in 1912. See "Extension Work among Negroes, 1920," U.S. Department of Agriculture, Department Circular 190 (Washington: Government Printing Office, 1921), 7.

6 Monroe N. Work, *Industrial Work of Tuskegee Graduates and Former Students during the Year 1910* (Tuskegee: Institute Press, 1911), 24; Booker T. Washington, "The Successful Training of the Negro: The Results of Such Teaching as Is Given at Tuskegee Institute-Schools Founded on the Tuskegee Idea," *Educational Number: The World's Work*, August 1903: 3750; U.S Department of Agriculture, "List of Workers in Subjects Pertaining to Agriculture and Home Economics in the U.S. Department of Agriculture and in the State Agricultural Colleges and Experiment Stations" (Washington: Government Printing Office, 1914), 55.

7 Gary Zellar, "H. C. Ray and Racial Politics in the African American Extension Service Program in Arkansas, 1915–1929," *Agricultural History* 72, no. 2 (Spring 1999): 431, 433; Mena Hogan, "A History of the Agricultural Extension Service in Arkansas," master's thesis, University of Arkansas, 1940, 139; Mary L. Ray death certificate, Arkansas Department of Health; Ray's funeral services were held at Bethel A.M.E. Church in Little Rock. She was survived by her husband and some cousins. See "Arkansas State," *Chicago Defender*, August 11, 1934, 21; University of Arkansas Libraries, Special Collections, University of Arkansas Cooperative Extension Service Records, 1914–1988, MC 1145, 2, 1, box 8, file 3. See also T. R. Betton, "A Brief History of Agricultural Extension Work among Negroes in Arkansas," 1962. H. C. Ray retired in 1952.

8 Carmen Harris, "Grace under Pressure: The Black Home Extension Service in South Carolina, 1919–1966," in *Rethinking Home Economics: Women and the History of a Profession*, ed. Sarah Stage (Ithaca: Cornell University Press, 1997), 207.

9 Hogan, "A History," 155.

10 Whayne, *Arkansas*, 309.

11 Helen Zoe Veit, *Modern Food, Moral Food: Self-Control, Science, and the Rise of Modern American Eating in the Early Twentieth Century* (Chapel Hill: University of North Carolina Press, 2013), 104.

12 Zellar, "H. C. Ray and Racial Politics," 433; Whayne, *Arkansas*, 309; Hogan, "A History," 147.

13 Betton, "A Brief History," 2.

14 Donald Holley, *Uncle Sam's Farmers: The New Deal Communities in the Lower Mississippi Valley* (Urbana: University of Illinois Press, 1975), 134.

15 Hogan, "A History," 141.

16 Veit, *Modern Food, Moral Food*, 115.

17 Hogan, "A History," 147, 148.

18 Peter Coclanis and Bryant Simon, "'Exit, Voice, and Loyalty,' African American Strategies for Day to Day Existence/Resistance in the Early Twentieth Century

Rural South," in *African American Life in the Rural South, 1900–1950*, ed. R. Douglas Hurt (Columbia: University of Missouri Press, 2003), 203.

19 For more information, see Isabel Wilkerson's Pulitzer Prize–winning, *The Warmth of Other Suns: The Epic Story of America's Great Migration* (New York: Random House, 2010), 8–9.

20 Betton, "A Brief History," 3; Whayne, *Arkansas*, 309; Nan Woodruff, *American Congo: The African American Freedom Struggle in the Delta* (Chapel Hill: University of North Carolina Press, 2003), 146.

21 Although most Jeanes teachers were female, this was not necessarily required. In Arkansas, some of the earliest Jeanes teachers were male. See Atlanta University Center, Robert W. Woodruff Library, Atlanta, Georgia, Archives Research Center, Southern Education Foundation Archives, box 66, folder 3, "Arkansas Jeanes Teachers"; Mildred M. Williams et. al., *The Jeanes Story: A Chapter in the History of American Education, 1908–1968* (Atlanta: Southern Education Foundation, 1979), 115–16.

22 "Farmers in Conference at Tuskegee Institute," *Chicago Defender*, February 7, 1925, 3.

23 B. D. Mayberry, *The Role of Tuskegee University in the Origin, Growth and Development of the Negro Cooperative Extension System, 1881–1990* (Tuskegee: Tuskegee University Cooperative Extension Program, 1989), 88.

24 "Many Win Prizes at Arkansas State Fair," *Chicago Defender*, November 22, 1924, A1; Blain Roberts, *Pageants, Parlors, and Pretty Women: Race and Beauty in the Twentieth Century South* (Chapel Hill: University of North Carolina Press, 2014), 118.

25 Donald Holley, "The Plantation Heritage: Agriculture in the Arkansas Delta," in *The Arkansas Delta: Land of Paradox*, ed. Jeannie Whayne and Willard B. Gatewood (Fayetteville: University of Arkansas Press, 1993), 260.

26 Pete Daniel, *Deep'n as It Come: The 1927 Mississippi River Flood* (Fayetteville: University of Arkansas Press, 1996), 36, 144. H. C. Ray was appointed an assistant to the State Reconstruction officer in Arkansas. See "The Final Report of the Colored Advisory Commission Appointed to Cooperate with the American National Red Cross and the President's Committee on Relief Work in the Mississippi Valley Flood Disaster of 1927" (Washington, DC: American Red Cross, 1929), 23.

27 Arkansas Extension Homemakers Council, *History of Home Demonstration Work in Arkansas 1914–1965: Extension Homemaker Work, 1966–1975* (n.p., 1978), 325.

28 "Arkansas Demonstration Agent Resigns Post," *Broad Axe*, Chicago, Illinois, September 10, 1927, 2.

29 "Memphis, Tenn," *Plaindealer* (Kansas City, Kansas), August 25, 1939, vol. 41, no. 32.5; "State Choir Gives Christmas Concert," *Arkansas State Press*, December 24, 1948, 1; Floyd Brown, *He Built a School with Two Dollars and Eighty-Five Cents: Floyd Brown, Founder, Fargo Agricultural School, Fargo, Arkansas* (Fargo, AR: 1956), 57. Toler established the home in 1942 and was its executive director. It was also known as the Lula Toler Convalescing Home. Lula Toler died on August 17, 1958. See *Toler v. Lula Toler Convalescing Home*, 364 S.W.2d 680 (1963), www.law.justia.com; "African American Firsts," *Pine Bluff Commercial*, www.pbcommercial.com; "Former PB Resident Protest Killing" *Pine Bluff Commercial*, www.pbcommercial.com.

30 Daniel, *Deep'n as It Come*, 36, 144.

31 National Archives and Records Administration-Southwest Region, Fort Worth, TX., Records of the Extension Service Record Group 33 (hereafter cited as NASW, RG 33), Narrative Report of Local Home Demonstration Agent, St. Francis County, Arkansas, 1927.

32 Martin Nelson, "Rural Hygiene," Extension Circular no. 31, June 1914 (Fayetteville: University of Arkansas Agricultural Experiment Station and College of Agriculture), n.p.

33 NASW, RG 33, Narrative Report of Local Home Demonstration Agent, Lee County, 1927.

34 Elizabeth Griffin Hill, *A Splendid Piece of Work: One Hundred Years of Arkansas' Home Demonstration and Extension Homemakers Clubs* (CreateSpace Independent Publishing Platform, 2012), 85. Symptoms of pellagra include red and scaly skin, stomach disorders, diarrhea, and depression. The U.S. Public Health Service conducted investigations of the disease, and two wealthy Northern philanthropists donated $15,000 to study pellagra. See Susan L. Smith, *Sick and Tired of Being Sick and Tired: Black Women's Health Activism in America, 1890–1950* (Philadelphia: University of Pennsylvania Press, 1995), 9; Whayne, *Arkansas*, 296.

35 NASW, RG 33, Narrative Report of the Negro County Home Demonstration Agent, Lee County, 1927.

36 Ibid.

37 "Coolidge Projects Inland Waterways for Flood Control," *New York Times*, October 4, 1927, 1; Whayne, *Arkansas*, 314; Zellar, "H. C. Ray and Racial Politics," 441–42; Daniel, *Deep'n as It Come*, 137, 144.

38 "Movable Schools Bring Modern Farm Methods to Rural Homes," *Chicago Defender*, September 20, 1924, 13; "Movable Schools," *New York Amsterdam News*, December 9, 1925, 10; Betton, "A Brief History," 3–4; Karen J. Ferguson, "Caught in 'No Man's Land': The Negro Cooperative Demonstration Service and the Ideology of Booker T. Washington, 1900–1918," *Agricultural History* 72, no. 1 (Winter 1998): 34–37.

39 Hogan, "A History," 149; Zellar, "H. C. Ray and Racial Politics," 444.

40 "Tells How Famous Movable School Started," *Chicago Defender*, October 12, 1940, 2; Lynne A. Rieff, "'Go Ahead and Do All You Can': Southern Progressives and Alabama Home Demonstration Clubs, 1914–1940," in *Southern Women: Hidden Histories of Women in the New South*, ed. Virginia Bernhard, Betty Brandon, Theda Purdue, and Elizabeth Turner (Columbia: University of Missouri Press, 1994), 139; Zellar, "H. C. Ray and Racial Politics," 444; Allen W. Jones, "The Role of Tuskegee Institute in the Education of Black Farmers," *Journal of Negro History* 60 (April 1975): 262–65.

41 Rieff, 136–37; Eunice Heywood, "Home Demonstration Clubs and Council," in *The Cooperative Extension Service*, ed. H. C. Sanders (Englewood Cliffs, N. J.: Prentice-Hall, 1966), 253.

42 Narrative Report of the Negro County Home Demonstration Agent, Woodruff, Arkansas, 1944, RG 33, reel 106, NASW; Ben F. Johnson III, *Arkansas in Modern America, 1930–1999* (Fayetteville: University of Arkansas Press, 2000), 34.

43 Linda Ambrose also addresses rural women's political empowerment in *Wom-*

en's Institutes in Canada: The First One Hundred Years, 1897–1997 (Ottawa: Tri-Co, 2000). See also her chapter "Forever Lunching" in this volume.

44 Mrs. J. Howard Crawford, President, Arkansas Council of Home Demonstration Clubs, Arkadelphia, "Home Demonstration Work in Arkansas," *Proceedings of the Arkansas Academy of Science* 4 (1951): 183.

45 Arkansas Extension Homemakers Council, *History of Home Demonstration Work*, 311; "Arkansas Farmers in 16th Annual Meet," *Chicago Defender*, September 5, 1936, 15; Hill, *A Splendid Piece of Work*, 260.

46 T. M. Campbell, "Home Demonstration Work among Negroes in South Has Aided Many," *Atlanta Daily World*, February 6, 1935, 2; Barbara R. Cotton, *The Lamplighters: Black Farm and Home Demonstration Agents in Florida, 1915–1965* (Tallahassee: U.S. Department of Agriculture in cooperation with Florida Agricultural and Mechanical University, 1982), 53; Melissa Walker, *All We Knew Was to Farm: Rural Women in the Upcountry South, 1919–1941* (Baltimore: Johns Hopkins University Press, 2000), 113.

47 Arkansas State Emergency Relief Administration, *Traveling Recovery Road: The Story of Relief, Work-Relief and Rehabilitation in Arkansas, August 30, 1932 to November 15, 1936* (Little Rock, Arkansas, 1936), 93–94.

48 Johnson, *Arkansas in Modern America*, 34.

49 MRS. J. B. WATSON Special, "GIRL 16, SUPERVISES CANNING KITCHEN," *Chicago Defender*, June 15, 1935, 15.

50 T. M. Campbell, "Home Demonstration Work"; Barbara R. Cotton, *The Lamplighters*; Walker, 113.

51 Carmen Harris, "'You're Just Like Mules, You Don't Know Your Own Strength': Rural South Carolina Blacks and the Emergence of the Civil Rights Struggle," in *Beyond Forty Acres and a Mule: African American Landowning Families since Reconstruction*, ed. Debra A. Reid and Evan P. Bennett (Gainesville: University Press of Florida, 2014), 256.

52 Evelyn Brooks Higginbotham, *Righteous Discontent: The Women's Movement in the Black Baptist Church, 1880–1920* (Cambridge, MA: Harvard University Press, 1993), 186–87.

53 Lu Ann Jones, *Mama Learned Us to Work: Farm Women in the New South* (Chapel Hill: University of North Carolina Press, 2002), 153.

54 "Arkansas Women Stage 'Cotton Dress' Revue," *Atlanta Daily World*, June 10, 1940, 2. See Hogan, "A History," 227, 232.

55 Janet Anne Hutchison, "American Housing, Gender, and the Better Homes Movement, 1922–1935," PhD diss., University of Delaware, 1989, 6. *The Delineator* was a woman's magazine owned by the Butterick Publishing Company.

56 "Better Homes in America: A Guide for Better Homes Campaigns in Rural Communities and Small Towns, Better Homes Week, April 21 to April 27, 1929," publication no. 17, September 1928, 22 and 24.

57 "Colored Teachers Close Institute," *Camden News* (Camden, Arkansas), June 20, 1930, 9.

58 Mary S. Hoffschwelle, *Rebuilding the Rural South Community: Reformers, Schools,*

and Homes in Tennessee, 1900–1930 (Knoxville: University of Tennessee Press, 1998), 137–38.

59 Michele Mitchell, *Righteous Propagation: African Americans and the Politics of Racial Destiny after Reconstruction* (Chapel Hill: University of North Carolina Press, 2004), 142.

60 Narrative Report of Negro Local Home Demonstration Agent, St. Francis County, 1933, RG 33, NASW.

61 "Arkansas Women Aid Better Homes Movement There," *Atlanta Daily World*, July 23, 1939, A6.

62 "McGehee Observes Better Homes Week," *Arkansas State Press*, November 9, 1941, 6.

63 "Elected Chairman of Better Homes Drive," *Arkansas State Press*, February, 5, 1943, 3; "Two Agricultural Employees Receive Superior Awards," *Atlanta Daily World*, June 9, 1956, 1; "Down Our Way," *Arkansas State Press*, January 7, 1944, 8.

64 "Dixie Sleeps Well, Due to U.S. Mattress Projects," *Chicago Defender*, February 21, 1942, 6; NASW, RG 33, reel 104, Narrative Report of the Negro County Home Demonstration Agent, Monroe County, 1944.

65 Arkansas Extension Homemakers Council, *History of Home Demonstration Work*, 328.

66 Interview with Delores Twillie Woods and Thelma Woods Nash (btvct02023), interviewed by Mausiki S. Scales, Forrest City (Arkansas), July 19, 1995, in *Behind the Veil: Documenting African-American Life in the Jim Crow South Digital Collection*, John Hope Franklin Research Center, Duke University Libraries. They also made sheets, pillowcases, and face towels out of flour sacks.

67 Roberts, *Pageants, Parlors, and Pretty Women*, 11.

68 Mitchell, *Righteous Propagation*, 148.

69 "Negro Farmer Confab Slated for Arkansas," *Chicago Defender*, July 4, 1942, 5; "Negro Farmers Part in Victory Drive to be Thrashed Out Soon," *Atlanta Daily World*, July 6, 1942, 5; "800 Attend AAA Meets in Arkansas," *Atlanta Daily World*, July 22, 1942, 2. It was also known as the "Food for Victory Campaign."

70 Ibid; T. R. Betton, "On the Agricultural Front with the U.S. Department of Agriculture," *Arkansas State Press*, April 13, 1945, 3; Veit, *Modern Food, Moral Food*, 114.

71 Edmund de Schweinitz Brunner and E. Hsin Pao Yang, *Rural America and the Extension Service* (Menasha, WI: George Banta, 1949), 87.

72 Pamela Webb, "By the Sweat of the Brow: The Back-to-the-Land Movement in Depression Arkansas," *Arkansas Historical Quarterly* 42, no. 4 (Winter 1983): 338; Charles Morrow Wilson, editorial correspondence, "Arkansas Moves to Guard Future," *New York Times*, February 15, 1931, 56; "Arkansas Farmers Win," *Atlanta Constitution*, November 1, 1931, 4C; Earle E. Griggs, "Backbone and Spareribs," *Atlanta Constitution*, November 1, 1931, 5C; photograph of Lena H. Eddington, *Arkansas State Press*, June 4, 1954, 1. In 1954, Eddington was the home demonstration agent for Pulaski County.

73 Harris, "Grace under Pressure," 217; June L. Rhue, "Arkansas H-D Agent Finds Meals Balanced (2)," *Chicago Defender*, October 6, 1951, 10; Brunner and Yang, *Rural America and the Extension Service*, 87; Johnson, *Arkansas in Modern America*, 34.

74 Joyce Kinkead, *Farm: A Multi-Modal Reader* (Southlake, TX: Fountainhead Press, 2014), 184.

75 NASW, RG 33, Report of Negro Local Home Demonstration Agent, Phillips County, 1936.

76 "Woodruff Co.," *Arkansas State Press*, February 10, 1950, 3.

77 The cypress "knees" are the part of the tree that is above water. See "Lake Village," *Arkansas State Press*, February 2, 1955, 4.

78 "Dermott," *Arkansas State Press*, March 2, 1955, 4.

79 "Cash Crops Help Small Ark[ansas] Farmer Make Good Living," *Chicago Defender*, December 22, 1956, 10.

80 "250 Arkansas Farm Families Lift Themselves by Their Own Bootstraps," *Arkansas State Press*, July 31, 1959, 7.

81 University of Arkansas Libraries, Special Collections, University of Arkansas Cooperative Extension Service Records, 1914–1988, MC 1145, 2, 1, box 8, file 3, Sixteenth Annual Meeting of the Arkansas Negro State Home Demonstration Council, September 11–12, 1952; Twenty-Third Annual Meeting of the Arkansas Negro State Home Demonstration Council, September 2–4, 1959; Twenty-Fifth Annual Meeting of the Arkansas Negro State Home Demonstration Council, August 30–September 1961.

82 Debra Reid, *Reaping a Greater Harvest* (College Station: Texas A and M University Press, 2007), 63; Roscoe C. Brown, D.D.S., "The National Negro Health Week Movement," *Journal of Negro Education* 6, no. 3 (July 1937): 553; "National Negro Health Week to Be Observed April 1 to 8, 1928," *Public Health Reports* 43, no. 8 (February 24, 1928): 457; Edward H. Beardsley, *A History of Neglect: Health Care for Blacks and Mill Workers in the Twentieth Century South* (Knoxville: University of Tennessee Press, 1987), 102.

83 "Health Week Is Factor in Race Relations," April 1, 1926, *National Notes* (Kansas City, Missouri) 28, no. 7, 10.

84 Smith, *Sick and Tired of Being Sick and Tired*, 33; Beardsley, *A History of Neglect*, 103.

85 Smith, *Sick and Tired*, 13. The Tuskegee program was actually based on a black health program begun two years earlier by the Negro Organization Society of Virginia. See Beardsley, *A History of Neglect*, 102; Smith, *Sick and Tired*, 36.

86 "150 Negroes Examined at Clinic Last Week," *Courier News* (Blytheville, Arkansas), April 7, 1930, 1.

87 "National Negro Health Week," *Arkansas State Press* (Little Rock), April 10, 1942, 4.

88 "29th Observance of Negro Health Week Observed," *Arkansas State Press*, March 26, 1943, 1. Because young black men were rejected for military service because of illnesses related to poor nutrition, Eddington asserted that "health on the home front" equaled "victory on the war front." Report of the Negro Home Demonstration Agent, Poinsett County, 1944, RG 33, reel 104, NASW. See also Jeannie M. Whayne, "The Segregated Farm Program in Poinsett County, Arkansas," *Mississippi Quarterly* 45 (1992): 436; Woodruff, *American Congo*, 146.

89 "State HD Council Holds 17th Annual Meeting at ABC (Arkansas Baptist College)," *Arkansas State Press*, September 25, 1953, 1.

90 NASW, RG 33, Narrative Report of Assistant County Agent and Assistant Home Demonstration Agent for Negro Work, Union County, 1954.

91 Smith, *Sick and Tired*, 102. Rural blacks were not passive in this process. Poor and working-class African Americans were not "mere silent partners in the quest for health reform." Ibid., 2.

92 "Arkansas Town Takes a New Lease on Life," *Chicago Defender*, February 6, 1960, 15.

93 "Housekeeping Contest Set for Residents of Ivory Heights," *Camden News* (Camden, Arkansas), March 29, 1963, front page.

94 "Arkansas Town Takes a New Lease on Life."

95 Ibid.

96 "Faulkner County Extension News," *Arkansas State Press*, March 10, 1950, 7. Mastitis is an infection of the breast tissue. "Upchurch," *Arkansas State Press*, February 2, 1954, 6; "Lake Village," *Arkansas State Press*, June 10, 1955, 3.

97 "Extension Titles Are Changed," *Hope Star* (Hope, Arkansas), December 10, 1966, front page; Charline J. Warren, ed., *An Official History of National Extension Homemakers Council, Inc.: 1930–1990* (Burlington, KY: National Extension Homemakers Council, 1991), 44. In 1968, the organization became the Arkansas Cooperative Extension Service.

98 "1965 County Budget," *El Dorado Times* (El Dorado, Arkansas), January 2, 1965, 3.

99 "Quorum County Sets Expenses," *Hope Star*, November 23, 1966, front page.

100 Pete Daniel, *Dispossession: Discrimination against African American Farmers in the Age of Civil Rights* (Chapel Hill: University of North Carolina Press, 2015), 212.

101 Telephone interview with Carreather Banks Perry, Forrest City, Arkansas, January 26, 2015.

Chapter 9

This chapter was previously published as Linda M. Ambrose, "Forever Lunching: Food, Power and Politics in Ontario Women's Organizations," *Popular Culture Review* 13, no. 1 (January 2002). 7–17. Used with permission of the journal.

1 "Mainly for Women," *Sudbury Star*, June 8, 1942, 10.

2 Franca Iacovetta, Valerie J. Korinek, and Marlene Epp, eds., *Edible Histories, Cultural Politics: Toward a Canadian Food History* (Toronto: University of Toronto Press, 2012).

3 Roszika Parker, *The Subversive Stitch: Embroidery and the Making of the Feminine* (New York: Routledge, 1984). Thanks to my colleague, Janice Liedl, for introducing me to Parker's work.

4 Ibid., vi.

5 Ibid., 215.

6 Ontario, Department of Agriculture, *Annual Report of the Women's Institutes*, 1918, 122. Hereafter cited as *AR*. See also Linda M. Ambrose, "The Women's Institutes in Northern Ontario, 1905–1930: Imitators or Innovators?," in *Changing Lives: Women in Northern Ontario*, ed. Margaret Kechnie and Marge Reitsma-Street (Toronto: Dundurn, 1996), 354n5.

7 *AR* 1912, 90.

8 *AR* 1912, "Question Box," 42.

9 Mrs. A. Brown, Sr., Winterbourne, North Waterloo, *AR* 1911, 33.

10 Mrs. E. Richmond, "Lonely Women," *AR* 1907, 128.

11 Ambrose, "The Women's Institutes in Northern Ontario," 270–74.

12 *AR* 1928, 39.

13 Terry Crowley, "The Origins of Continuing Education for Women: The Ontario Women's Institutes," *Canadian Woman Studies* 7, no. 3 (Fall 1986): 78–81.

14 Author correspondence with the Canfield WI, printed material enclosed with survey response.

15 *AR* 1913, 30.

16 *AR* 1917, 30; *AR* 1919, 39–40.

17 *Home and Country*, May 1937, 1; *AR* 1936, 68.

18 *AR* 1952, 182.

19 Robert M. Stamp, "Teaching Girls Their 'God-Given Place in Life': The Introduction of Home Economics in the Schools," pt. 1, *Atlantis* 2, no. 2 (Spring 1977): 18–34.

20 Margaret Kechnie, *Organizing Rural Women: The Federated Women's Institutes of Ontario, 1897–1919* (Montreal and Kingston: McGill-Queen's University Press, 2003).

21 Ruby Heap, "From the Science of Housekeeping to the Science of Nutrition: Pioneers in Canadian Nutrition and Dietetics at the University of Toronto's Faculty of Household Science, 1900–1950," in *Challenging Professions: Historical and Contemporary Perspectives on Women's Professional Work*, ed. Elizabeth Smyth, Sandra Acker, Paula Bourne, and Alison Prentice (Toronto: University of Toronto Press, 1999): 141–70.

22 Karen Dubinsky, "'Who Do You Think Did the Cooking?' Baba in the Classroom," in *Changing Lives*, 193–97.

23 Interview with Ivy White [Irwin], Callander, Ontario, 1994. See also Ambrose, *For Home and Country*, 123.

24 Dorothy H. Dobbie, "Mother's at the W.I.," correspondence with the author from Hilde G. Morden, 1994.

25 Joseph Hudspeth, "A Husband's View of the Women's Institute," Caledonia, Ontario, n.d. [c. 1915]. This poem was provided to the author in correspondence from Mrs. S. Chartier, Lyn Ontario, March 8, 1994.

26 Canadian women won the right to vote earlier than either British or American women. In most provinces voting rights were extended to women at various times between 1916 and 1925; in Quebec rights were not granted until 1940. Canadian women achieved the right to vote in federal elections in 1919, although the 1917 election had provided special measures that allowed some women to vote, including nurses serving in the war and "wives, widows, mothers, sisters, and daughters of

those, alive or deceased, who had served or were serving in the Canadian or British military or naval forces." Gail Cuthbert Brandt, Magda Fahrni, Naomi Black, and Paula Bourne, *Canadian Women: A History*, 3rd ed. (Toronto: Nelson Education, 2011), 269.

27 Gore Bay Public Library, Barrie Island WI, *Tweedsmuir History Book*.

28 Linda M. Ambrose, "'What Are the Good of Those Meetings Anyway?': Early Popularity of the Ontario Women's Institutes," *Ontario History* 57, no. 1 (Spring 1995): 5.

29 Stratton WI, *Tweedsmuir History Book*, private collection.

30 Ibid.

31 *AR* 1924, 32.

32 Carol J. Dennison, "'Housekeepers of the Community': The British Columbia Women's Institutes, 1909–1946," in *Knowledge for the People: The Struggle for Adult Learning in English-Speaking Canada, 1828–1973*, ed. Michael R. Welton (Toronto: OISE, 1987), 52–72.

33 Mrs. W. Buchanan, Ravenna, "Address to the Central Ontario WI Convention, 1915," *AR* 1916, 166.

Chapter 10

1 Bureau of American Ethnology Records, ser. 1, Correspondence, Letters Received, 1909–1949, box 152, Smithsonian Institution, National Anthropology Archives, Washington, DC. Hereafter cited as Densmore-Hodge BAE Correspondence.

2 Discussions and comments by anthropologists and ethnologists were a crucial part of research on indigenous knowledge. For overviews of the broad range of issues involved in obtaining and exploiting this knowledge by nonnative peoples, see Vanda Shiva, "Foreword: Cultural Diversity and the Politics of Knowledge," in *Indigenous Knowledges in Global Contexts: Multiple Readings of Our World*, ed. George J. Sefa Dei, Budd L. Hall, and Dorothy Goldin Rosenberg (Toronto: University of Toronto Press, 2000), vii–x; and Ellen Simmons, *Indigenous Earth: Praxis and Transformation* (Penticon, BC: Theyfus, 2010). The most recent study of Frances Densmore is Joan M. Jen sen and Michelle Wick Patterson, eds., *Travels with Frances Densmore: Her Life, Work, and Legacy in Native American Studies* (Lincoln: University of Nebraska Press, 2015).

3 See Brenda J. Childs's introduction to Frances Densmore, *Strength of the Earth: The Classic Guide to Ojibwe Uses of Native Plants* (St. Paul: Minnesota Historical Society Press, 2005). For the practice of collaboration, see Luke Eric Lassiter, *The Chicago Guide to Collaborative Ethnography* (Chicago: University of Chicago Press, 2005). A major exception to the exclusion of gender discussions was Patricia L. Howard, ed., *Women and Plants: Gender Relations in Biodiversity Management and Conservation* (London: Zen, 2003), which contains Nancy J. Turner, "'Passing the News': Women's Work, Traditional Knowledge, and Plant Resource Management in Indigenous Societies of

Northwestern North America," 133–49. For the history of ethnobotany, see Richard Evans Schultes and Siri von Reis, eds., *Ethnobotany: Evolution of a Discipline* (Portland: Dioscorides, 1997), which includes Nancy J. Turner's, "Ethnobotany Today in Northwestern North America," 264–83. Turner has used Densmore's Makah botany extensively in her compilations of ethnobotanical reports on the Northwest.

4 James L. Reveal, *Gentle Conquest: The Botanical Discovery of North America* (Washington, DC: Starwood, 1992); Peter A. Zakrzeswki, "Bioprospecting or Biopiracy? The Pharmaceutical Industry's Use of Indigenous Medicinal Plants as a Source of Potential Drug Candidates," *University of Toronto Medical Journal* 79, no. 3 (May 2002): 252–54. See also Vandana Shiva, *Biopiracy: The Plunder of Nature and Knowledge* (Cambridge, MA: South End Press, 1999); and, for collaboration with a rural sociologist, Cornelia Butler Flora, Iva Honyestewa, and Harrissa Koiyaquaptewa, "Understanding Access and Use of Traditional Food by Hopi Women," *Journal of Hunger and Environmental Nutrition* 4 (2009): 158–71.

5 Everett E. Edwards and Wayne Rasmussen, comps., *A Bibliography on the Agriculture of the American Indians*, U.S. Department of Agriculture, Miscellaneous Publication no. 447 (Washington, DC: Government Printing Office, 1942); quotation from the earlier mimeographed "Bibliography," USDA Library Bibliographic Contributions 23, 1st ed. 1, (May 1932; 2nd ed., June 1933), 1.

6 Turner, "Ethnobotany Today in Northwestern North America," 266.

7 A. C. Parker, *Iroquois Uses of Maize and Other Food Plants*, New York State Museum, Museum Bulletin 144 (Albany: University of State of New York, 1910), 18–20.

8 See Lawrence C. Kelly, ed. *Navajo Roundup: Selected Correspondence of Kit Carson's Expedition against the Navajo, 1863–1865* (Boulder: Pruett, 1970).

9 See, for example, Ronald N. Satz, *Chippewa Treaty Rights: The Reserved Rights of Wisconsin's Chippewa Indians in Historical Perspective*, Wisconsin Academy of Sciences, Arts and Letters, *Transactions* 79, no. 1 (1991). Kevin Gover, "Nation to Nation: Treaties between the United States and American Indian Nations," *National Museum of the American Indian* 15, no. 2 (Summer/Fall 2014): 37–52, discusses the museum's exhibit on the history of treaty rights.

10 For the widespread practice of feasting, see Barbara J. Mills, *Identity, Feasting, and the Archaeology of the Greater Southwest* (Boulder: University Press of Boulder, 2004).

11 Joan M. Jensen, "The Mescalero Basketmakers," in *Out of the Shadows: The Women of Southern New Mexico*, ed. Martha Shipman Andrews (Los Ranchos: Rio Grande Books, 2012), 89.

12 Andrea Wulf, *Founding Gardeners: The Revolutionary Generation, Nature, and the Shaping of the American Nation* (New York: Knopf, 2011), 154–72.

13 For the six entries and four plates, see Gary E. Mouton, ed., *Herbarium of the Lewis and Clark Expedition* (Lincoln: University of Nebraska Press, 1999), vol. 12 of *The Journals of the Lewis and Clark Expedition* (Lincoln: University of Nebraska Press, 1983–1999), 25, 34–36, 49.

14 For the 1838–1842 expedition and the reign of William R. Smith, see Anne-

Catherine Fallen, *A Botanic Garden for the Nation: The United States Botanic Garden* (Washington, DC: United States Botanic Garden, 2007), 20–22, 29; see also Reveal, *Gentle Conquest*, 37.

15 John Bartlett, *Personal Narrative of Explorations and Incidents in Texas, New Mexico, California, and Chihuahua* (New York: Appleton, 1854), 1:179, 234.

16 Frank Hamilton Cushing, *Zuni Breadstuff*, Indian Notes and Monographs, vol. 8 (New York: Museum of the American Indian, 1920; repr., Nabu Press, 2014); *The Sacred Formulas of the Cherokees*, Seventh Annual Report of the Bureau of Ethnology to the Secretary of the Smithsonian Institution, 1885–1886 (Washington, DC.: Government Printing Office, 1891), 301–98, repr. in James Mooney's *History, Myths, and Sacred Formulas of the Cherokees* (Fairview, NC: Bright Mountain, 1992). See Virgin J. Vogel, *American Indian Medicine* (Norman: University of Oklahoma Press, 1990); a reprint of his 1970 edition with a new preface traces the early interest of the BAE in indigenous medicine to the late 1880s and the work reported by James Mooney.

17 See Edwin Palmer, Ian W. Brown, and Marvin D. Jeter, *Edward Palmer's Arkansas Mounds* (Fayetteville: University of Arkansas Press, 1990).

18 John Harshberger, "The Purposes of Ethno-botany," *American Antiquarian* 17 (1896): 73–81. He had done no research on the process of growing of corn but inherited corn specimens from the 1893 Chicago World's Fair, identified only as collected by Wetherill and by a person named Hazzard. He coined the term "ethno-botany" and urged its study. David Prescott Barrows, *The Ethno-Botany of the Coahuilla Indians of Southern California* (Chicago: University of Chicago Press, 1900), a dissertation completed in 1897, was probably the first academic study to use "ethnobotany" in a title and attempt to trace its history. The BAE had, however, published Washington Mathews's work on Navajo medicine in in its 1883–1884 annual report, 379–467, and James Mooney's "The Sacred Formulas of the Cherokees," in its 1885–86 annual report, 301–97.

19 Patterson, "She Always Said," *Travels with Frances Densmore*, 40–41.

20 Ibid., 41.

21 Matilda Coxe Stevenson, *The Zuni Indians and Their Uses of Plants*, U.S. Bureau of American Ethnology Annual Report (1908–9), 31–102.

22 Patterson, "She Always Said," 41.

23 See Densmore-Hodge BAE Correspondence, F. W. Hodge to F. V. Coville, January 10, 1914, asking for help in identifying plants.

24 Densmore-Hodge BAE Correspondence, Densmore to Hodge, June 25, 1917.

25 Michael D. McNally, *Ojibwe Singers: Hymns, Grief, and a Native Culture in Motion* (St. Paul: Minnesota Historical Society Press, 2000), 134–36, including a photograph of the brass band.

26 The only photograph of English in the Minnesota Historical Society archive is dated 1920, when she would have been eighty-five years old. See Frances Densmore, *Chippewa Customs*, Bureau of American Ethnology, Bulletin 86 (Washington, DC: Government Printing Office, 1929); *Chippewa Music*, Bureau of American Ethnology, Bulletin 45 (Washington, DC: Government Printing Office, 1910); and *Chippewa*

Music, Bureau of American Ethnology, Bulletin 53 (Washington, DC: Government Printing Office, 1913), 8.

27 Densmore, *Chippewa Customs*, 4–5.

28 See W. Roger Buffalohead, introduction to William Warren, *History of the Ojibway People* (St. Paul: Minnesota Historical Society Press, 1984); see also Theresa Schenck's introduction to the 2009 Minnesota Historical Society reprint, which cites an undated Mary Warren English manuscript in which she discusses William's methods.

29 Densmore, *Chippewa Customs*, 4–5.

30 For early mixed heritage communities, see Lucy Eldersveld Murphy, *Great Lakes Creoles: A French-Indian Community on the Northern Borderlands, Prairie du Chien, 1750–1850* (New York: Cambridge University Press, 2014).

31 Patterson, "She Always Said," 29–64.

32 Episcopal Bishop Whipple was well known for his support of Native American rights, and the Beaulieu family was prominent in the business and social life of White Earth. Thus, Episcopalian contacts were important locally. See Bruce White, *We Are at Home: Pictures of the Ojibwe People* (St. Paul: Minnesota Historical Society Press, 2007), 89–94.

33 Densmore discussed the specimens' use in *Collection of Specimens from the Teton Sioux: Indian Notes and Monographs* 11, no. 3 (New York: Museum of the American Indian Heye Foundation, 1948), 177–80. Densmore mentions only yarrow and notes that healers kept a large supply of herbs in their homes but carried only a small amount in their pouches. See also Densmore's *Teton Sioux*, 253–67. Her 1910 *Chippewa Music*, 51, 119, contains only brief comments on herbs used with "medicine" songs. See also Densmore-Hodge BAE Correspondence, Frances Densmore to Frederick Hodge, November 14, 1912.

34 Densmore-Hodge BAE Correspondence, box 152, Frances Densmore to Frederick Hodge, June 25, 1917.

35 Ibid., Frances Densmore to Frederick Hodge, July 2, 1917; August 10, 1917; August 26, 1917; August 28, 1917; September 8, 1917; September 11, 1917.

36 Ibid., Frances Densmore to Frederick Hodge, December 15, 1921.

37 Densmore-Fewkes BAE Correspondence, box 153. Densmore to J. W. Fewkes, July 1, 1921. For a more detailed description of White Earth during these years, see Melissa L. Meyer, *The White Earth Tragedy: Ethnicity and Dispossession at a Minnesota Anishinaabe Reservation* (Lincoln: University of Nebraska Press, 1994).

38 Michelle Wick Patterson, *Natalie Curtis Burlin: A Life in Native and African American Music* (Lincoln: University of Nebraska Press, 2010), 190, mentions that Burlin hoped Densmore would attend her Washington lecture in December 1908.

39 National Archives of Anthropology, Records of the Bureau of American Ethnology, box 153, Frances Densmore to Fewkes, January 19, 1926. For Densmore's comments on healing, see her "The Use of Music in the Treatment of the Sick by American Indians," *Musical Quarterly* 13 (January 4, 1928), 555–65. Reprinted in Charles Hoffman, *Frances Densmore and American Indian Music: A Memorial Volume* (New York:

Museum of the American Indian, Heye Foundation, 1968), 67–76; comment about "conquering nature," 68.

40 Joan M. Jensen, *Calling This Place Home: Women on the Wisconsin Frontier, 1850–1925* (St. Paul: Minnesota Historical Society, 2005), 168–69 (summarizing her descriptions).

Contributors

LINDA M. AMBROSE is a professor of history at Laurentian University in Sudbury, Ontario. Her research on rural women has focused on women's organizations, particularly the Women's Institutes. Her books include *For Home and Country: The Centennial History of the Women's Institutes in Ontario, 1897–1997* and *A Great Rural Sisterhood: Madge Robertson Watt and the ACWW* (University of Toronto Press, 2015). Ambrose's other research interest is the history of twentieth-century religion, and she is currently using biography to explore the gendered history of Canadian Pentecostalism; she is coeditor of the *Canadian Journal of Pentecostal-Charismatic Christianity*.

MAGGIE ANDREWS is a professor of cultural history at the University of Worcester, UK. Her research and publications have explored the interrelationship of domesticity and femininity and most recently the home front in World War I and II in Britain. Her current work with the website *Voices of War and Peace: The Great War and Its Legacy First World War Engagement Centre* is funded by the Arts and Humanities Research Council. Maggie is the author of *The Acceptable Face of Feminism: The Women's Institutes as a Social Movement*, first published in 1997 and released in a new and revised edition in 2015.

MARGREET VAN DER BURG is a senior university lecturer and researcher for gender in food, agricultural and rural studies at Wageningen University, Netherlands. In 2006 she was a guest professor at Georg August University, Göttingen, Germany, and in 2015–16 a visiting researcher at Pennsylvania State University. Her research focuses on farm women and gender in rural and agricultural development processes, especially agricultural knowledge systems, labor and care arrangements in family-based farming systems and agrifood chains; food security and resource allocation; governance and self-organization; and rurality in health and food movements, rural iconography, and cultural heritage.

JOAN M. JENSEN is a professor emerita at New Mexico State University, where she taught from 1976 to 1993. She founded the women's studies program at NMSU and was instrumental in the establishment of the Rural Women's Studies Association. Jensen has published widely on rural women's history; her books include *With These Hands: Women Working on the Land; Promise to the Land: Essays on Rural Women,* the Pulitzer Prize–nominated *Loosening the Bonds: Mid-Atlantic Farm Women, 1750–1850,* and *Calling This Place Home: Women on the Wisconsin Frontier, 1850–1925.*

CHERISSE JONES-BRANCH is a full professor of history at Arkansas State University–Jonesboro, where she teaches courses in United States, women's studies, civil rights, rural history, African American history, and heritage studies. She is the author of many articles on women's civil rights activism. In 2014, she published *Crossing the Line: Women and Interracial Activism in South Carolina during and after World War II* (University Press of Florida) and is coeditor of *Arkansas Women: Their Lives and Times* (forthcoming, University of Georgia Press). She is also working on a monograph on rural black women's activism in Arkansas.

AMY L. MCKINNEY is an associate professor of history at Northwest College in Powell, Wyoming. She completed her doctorate at the University of Calgary in 2011 and her masters at Montana State University in Bozeman in 2000. Her dissertation, "'How I Cook, Keep House, Help with Farm Work Too': Rural Women in Post–World War II Montana," examines how and to what degree the suburban housewife ideal translated to rural areas in Montana after World War II. She has presented and published several works on rural women. McKinney's current work investigates Harriette Cushman's role as a professional woman in agriculture.

ANNE L. MOORE is a special collections librarian in the special collections and university archives at the University of Massachusetts Amherst. She holds a certificate in archives management from Simmons College, an MLIS from the University of Western Ontario, and an MA in history from the University of Guelph, Ontario.

KAREN SAYER is a professor of social and cultural history at Leeds Trinity University, UK. She is the author of *Women of the Fields: Representations of Rural Women in the Nineteenth Century* (Manchester University Press, 1995) and *Country Cottages: A Cultural History* (Manchester University Press, 2000). She has published on egg production and consumption and is working on a monograph on the history of farm animals in modern Britain. Sayer's research focus is on British rural history and the conceptualizations of the rural, rurality, pastoral, and nature; human and animal relations in the countryside; and agricultural work and labor in field, home, farm, and dairy as represented and lived.

NICOLA VERDON is a reader in history at Sheffield Hallam University, UK. Her research focuses on the social and economic history of the British countryside in the nineteenth and twentieth centuries. She has published widely on gender and agricultural labor, farm families, and rural standards of living and is currently completing *Working the Land: A History of the Farm Worker in England from 1850 to the Present Day* (forthcoming, Palgrave).

Index

Note: Boldface indicates terms mentioned in captions